The Electric Force of a Current
Weber and the surface charges of
resistive conductors carrying steady currents

Andre Koch Torres Assis
Julio Akashi Hernandes

Apeiron
Montreal

Published by C. Roy Keys Inc.
4405, rue St-Dominique
Montreal, Quebec H2W 2B2 Canada
http://redshift.vif.com

© Andre Koch Torres Assis and Julio Akashi Hernandes. 2007

First Published 2007

Library and Archives Canada Cataloguing in Publication

Assis, André Koch Torres, 1962-
  The electric force of a current : Weber and the surface charges of resistive conductors carrying steady currents / Andre Koch Torres Assis, Julio Akashi Hernandes.

ISBN 978-0-9732911-5-5
Includes bibliographical references and index.

  1. Electric circuits. 2. Electric conductors. 3. Electrostatics. 4. Electromagnetism. I. Hernandes, Julio Akashi, 1977- II. Title.

QC610.4.A47 2007   537'.2   C2007-901366-X

Front cover: Portrait of Wilhelm Eduard Weber (1804-1891) around 1865. He was one of the pioneers of the study of surface charges in resistive conductors carrying steady currents.

Back cover: Figure of the 2 circuits: A constant current I flows along a resistive wire connected to a battery V. At the left side there is a qualitative representation of the charges along the surface of the wire. At the right side there is a representation of the internal and external electric fields generated by this distribution of surface charges.

Contents

Acknowledgments · iii

Foreword · v

Vorwort · vii

I Introduction · 1

1 Main Questions and False Answers · 7
1.1 Simple Questions · 7
1.2 Charge Neutrality of the Resistive Wire · 9
1.3 Magnetism as a Relativistic Effect · 13
1.4 Weber's Electrodynamics · 14
1.5 Electric field of Zeroth Order; Proportional to the Voltage of the Battery; and of Second Order · 20

2 Reasons for the Existence of the External Electric Field · 23
2.1 Bending a Wire · 23
2.2 Continuity of the Tangential Component of the Electric Field · 26

3 Experiments · 29
3.1 Zeroth Order Electric Field · 29
3.2 Electric Field Proportional to the Voltage of the Battery · 30
3.3 Second Order Electric Field · 42

4 Force Due to Electrostatic Induction · 45
4.1 Introduction · 45
    4.1.1 Point Charge and Infinite Plane · 45
    4.1.2 Point Charge and Spherical Shell · 46
4.2 Point Charge and Cylindrical Shell · 46
4.3 Finite Conducting Cylindrical Shell with Internal Point Charge: Solution of Poisson's Equation · 47
    4.3.1 Cylindrical Shell Held at Zero Potential · 49
4.4 Infinite Conducting Cylindrical Shell with Internal Point Charge · 50

		4.4.1 Cylindrical Shell Held at Zero Potential	50
	4.5	Infinite Conducting Cylindrical Shell with External Point Charge	52
		4.5.1 Cylindrical Shell Held at Zero Potential	53
		4.5.2 Thin Cylindrical Shell Held at Zero Potential	56
		4.5.3 Infinite Cylindrical Shell Held at Constant Potential . . .	58
	4.6	Discussion	61
5	**Relevant Topics**		**65**
	5.1	Properties of the Electrostatic Field	65
	5.2	The Electric Field in Different Points of the Cross-section of the Wire	66
	5.3	Electromotive Force Versus Potential Difference	67
	5.4	Russell's Theorem	68

II Straight Conductors   71

6	**A Long Straight Wire of Circular Cross-section**	**75**
	6.1 Configuration of the Problem	75
	6.2 Force Proportional to the Potential Difference Acting upon the Wire	77
	6.3 Force Proportional to the Square of the Current	82
	6.4 Radial Hall Effect	84
	6.5 Discussion	86
7	**Coaxial Cable**	**93**
	7.1 Introduction	93
	7.2 Potentials and Fields	94
	7.3 The Symmetrical Case	97
	7.4 The Asymmetrical Case	98
	7.5 Discussion	100
8	**Transmission Line**	**103**
	8.1 Introduction	103
	8.2 Two-Wire Transmission Line	103
	8.3 Discussion	108
9	**Resistive Plates**	**113**
	9.1 Introduction	113
	9.2 Single Plate	113
	9.3 Two Parallel Plates	116
	9.4 Four Parallel Plates	117
	9.4.1 Opposite Potentials	118
	9.4.2 Perfect Conductor Plate	120

10 Resistive Strip   **123**
   10.1 The Problem 123
   10.2 The Solution 124
   10.3 Discussion 126
   10.4 Comparison with the Experimental Results 128

III   Curved Conductors   133

11 Resistive Cylindrical Shell with Azimuthal Current   **137**
   11.1 Configuration of the Problem 137
   11.2 Potential and Electric Field 138
   11.3 Surface Charge Densities 141
   11.4 Representation in Fourier Series 143
   11.5 Lumped Resistor 146

12 Resistive Spherical Shell with Azimuthal Current   **151**
   12.1 Introduction 151
   12.2 Description of the Problem 151
   12.3 General Solution 153
   12.4 Electric Field and Surface Charges 156
   12.5 Conclusion 160

13 Resistive Toroidal Conductor with Azimuthal Current   **163**
   13.1 Introduction 163
   13.2 Description of the Problem 163
   13.3 General Solution 166
   13.4 Particular Solution for a Steady Azimuthal Current 167
   13.5 Potential in Particular Cases 170
   13.6 Electric Field and Surface Charges 173
   13.7 Thin Toroid Approximation 174
   13.8 Comparison of the Thin Toroid Carrying a Steady Current with the Case of a Straight Cylindrical Wire Carrying a Steady Current 180
   13.9 Charged Toroid without Current 181
   13.10 Comparison with Experimental Results 184

IV   Open Questions   189

14 Future Prospects   **191**

Appendices   **195**

A   Wilhelm Weber and Surface Charges   **195**

B   Gustav Kirchhoff and Surface Charges   **213**

Bibliography	**217**
Index	**236**

This book is dedicated to the memory of Wilhelm Eduard Weber (1804-1891). He was one of the main pioneers in the subject developed here, the study of surface charges in resistive conductors carrying steady currents. We hope this book will help to make his fundamental work better known.

Acknowledgments

The authors wish to thank many people who collaborated with them in previous works related to the topic of this book, and also several others for their support, advice, suggestions, references, etc. In particular they thank Waldyr A. Rodrigues Jr., A. Jamil Mania, Jorge I. Cisneros, Hector T. Silva, João E. Lamesa, Roberto A. Clemente, Ildefonso Harnisch V., Roberto d. A. Martins, A. M. Mansanares, Edmundo Capelas de Oliveira, Álvaro Vannucci, Iberê L. Caldas, Daniel Gardelli, Regina F. Avila, Guilherme F. Leal Ferreira, Marcelo de A. Bueno, Humberto de M. França, Roberto J. M. Covolan, Sérgio Gama, Haroldo F. de Campos Velho, Marcio A. d. F. Rosa, José Emílio Maiorino, Mark A. Heald, G. Galeczki, P. Graneau, N. Graneau, John D. Jackson, Oleg D. Jefimenko, Steve Hutcheon, Thomas E. Phipps Jr., J. Paul Wesley, Junichiro Fukai, J. Guala-Valverde, Howard Hayden, Hartwig Thim, D. F. Bartlett, F. Doran, C. Dulaney, Gudrun Wolfschmidt, Karin Reich, Karl H. Wiederkehr, Bruce Sherwood, Johann Marinsek, Eduardo Greaves, Samuel Doughty, H. Härtel and C. Roy Keys.

AKTA wishes to thank Hamburg University and the Alexander von Humboldt Foundation of Germany for a research fellowship on "Weber's law applied to electromagnetism and gravitation." This research was developed at the Institut für Geschichte der Naturwissenschaften (IGN) of Hamburg University, Germany, in the period from August 2001 to November 2002, during which he first had the idea to write this book. He was extremely well received in a friendly atmosphere and had full scientific and institutional support from Prof. Karin Reich and Dr. K. H. Wiederkehr. JAH wishes to thank CNPq, Brazil, for financial support. The authors thank also FAEP-UNICAMP for financial support to this project, and the Institute of Physics of the State University of Campinas - UNICAMP which provided them the necessary conditions to undertake the project.

A. K. T. Assis[*] and J. A. Hernandes[†]

---

[*]Institute of Physics, State University of Campinas, 13083-970 Campinas - SP, Brazil, E-mail: assis@ifi.unicamp.br, Homepage: http://www.ifi.unicamp.br/~assis

[†]Universidade Bandeirante de São Paulo - UNIBAN, São Paulo - SP, Brazil, E-mail: jahernandes@gmail.com

Foreword

Is there an interaction - some reciprocal force - between a current-carrying conductor and a stationary charge nearby? Beneath this simple question lie some remarkable misunderstandings, which are well illustrated by the fact that the answers to it commonly found in the scientific literature and also in many text books are incorrect.

In case there is any uncertainty about the answer, all doubt will be eliminated by this book. It tackles the question in a brilliant and comprehensive manner, with numerous hints for relevant experiments and with impressive mathematical thoroughness.

It is astonishing to learn that, as early as the middle of the 19th century, the German physicists Weber and Kirchhoff had derived and published the answer to this problem; however, their work was poorly received by the scientific community, and many rejected it as incorrect. The reasons behind this scientific setback, which are presented in detail in this book and supported with numerous quotations from the literature, represent a real treasure trove for readers interested in the history of science.

It becomes clear that even in the exact science of physics people at times violate basic scientific principles, for instance by referring to the results of experiments which have never been carried out for the purpose under discussion. This book helps readers not only to develop a detailed knowledge of a seriously neglected aspect of the so-called simple electric circuit, but reminds us also that even eminent physicists can be mistaken, that mistakes may be transferred from one textbook generation to the next and that therefore persistent, watchful and critical reflection is required.

A didactic comment is appropriate here. The traditional approach to teaching electric circuits based on current and potential difference is called into question by this book.

When dealing with electric current one usually pictures drifting electrons, while for the terms "voltage" or "potential difference" one directly refers to the abstract notion of energy, with no opportunity for visualization. Experience shows that only few school students really understand what "voltage" and "potential difference" mean. The inevitable result of failure to understand such basic terms is that many students lose interest in physics. Those whose confidence in their understanding of science is still fragile, may attribute failure to grasp these basic concepts as due to their own lack of talent.

Physics remains a popular and crucial subject, so the large numbers of students who each year study the subject implies that the search for less abstract and therefore more readily understood alternatives to traditional approaches is urgent.

This book offers such an alternative. It shows that in respect to surface charges there is no fundamental difference between an electrostatic system and the flow of an electric current. It refers to recent curriculum developments concerning "voltage" and "potential difference" and presents a comprehensive survey of related scientific publications, that have appeared since the early papers by Weber and Kirchhoff.

Why should we refer to drifting electrons when we teach electric current and yet not refer to drifting surface charges when teaching voltage or potential difference?

The final objective of the curriculum when voltage is covered will certainly be to define it quantitatively in terms of energy. For didactic reasons, however, it does not seem to be justifiable to omit a qualitative and more concrete preliminary stage, unless there is a lack of knowledge about the existence of surface charges. In the present market there are newly developed curriculum materials that cover basic electricity, to which the content of this book relates strongly. Comparison of the approach that this book proposes with more traditional approaches should dispel any doubts about the need for the methods that it describes.

This book provides a crucial step along the path to a better understanding of electrical phenomenon especially the movement of electrons in electic circuits.

Hermann Härtel
Guest scientist at Institut für Theoretische Physik und Astrophysik
Universität Kiel
Leibnizstrasse 15
D-24098 Kiel, Germany
E-mail: haertel@astrophysik.uni-kiel.de

Vorwort

Gibt es eine Wechselwirkung zwischen einem stromführenden Leiter und einem stationären Ladungsträger? Diese lapidare Frage enthält eine erstaunliche Brisanz, zumal die Antworten, die man bis zu diesem Tag in der Fachliteratur und auch in weit verbreitenden Lehrbüchern findet, häufig unzutreffend sind. Das vorliegende Buch beantwortet die Eingangsfrage in brillanter Weise: umfassend, mit zahlreichen Verweisen auf entsprechende Versuche und mit rigoroser, mathematischer Gründlichkeit.

Sofern Zweifel an einer positiven Antwort vorhanden waren, sind diese nach dem Studium des Buches ausgeräumt.

Erstaunlicherweise wurde bereits Mitte des 19 Jahrhunderts von den deutschen Physikern Weber und Kirchhoff eine zutreffende Antwort veröffentlicht, die jedoch von der wissenschaftlichen Gemeinde kaum rezipiert, teilweise sogar als unzutreffend zurückgewiesen wurde. Die Gründe für diesen wissenschaftlichen Rückschritt, die in dem Buch ausführlich dargestellt und mit zahlreichen Literaturzitaten belegt werden, stellen eine wahre Fundgrube für wissenschaftshistorisch interessierte Leser dar.

Sie machen deutlich, daß auch in der Physik als exakte Wissenschaft manchmal gegen methodische Grundprinzipien verstoßen wird, in dem zum Beispiel ein Verweis auf Experimente erfolgt, die nie gezielt durchgeführt wurden. So verhilft dies Buch seinen Lesern nicht nur zu einer fundierten Kenntnis über einen stark vernachlässigten Bereich des sogenannten einfachen elektrischen Stromkreises, sondern bringt in Erinnerung, daß auch die führenden Vertreter unserer Disziplin irren können, daß unter Umständen solche Irrtümer von einer Lehrbuchgeneration auf die nächste übertragen werden und somit beständige, wachsame und kritische Reflexion geboten ist.

Eine didaktische Anmerkung erscheint angebracht. Die im Physikunterricht übliche Vermittlung des elektrischen Stromkreis mit den Grundbegriffen Strom und Spannung, wird durch den Inhalt des vorliegenden Buches grundlegend in Frage gestellt.

Während zum Begriff des elektrischen Stromes noch Bilder von driftenden Elektronen angeboten werden, findet die Einführung der Spannung bzw. des Potentials auf der abstrakteren Ebene der Energie statt und läßt daher keinerlei Veranschaulichung zu. Wie die Erfahrung zeigt gelangen nur wenige Schüler zu ein tieferes Verständnis des Spannungsbegriffs. Dagegen führt bei vielen Schülern ein solches Scheitern gerade an einem so grundlegenden Begriff wie dem der Spannung zur Aufgabe des Interesses an physikalischen Inhalten. Vor allem jüngere Schüler mit noch schwach entwickeltem Selbstvertrauen mögen ein solches Scheitern sich selbst und dem eigenen Unvermögen zuschreiben?

Physik ist ein allgemein bildendendes und wichtiges Fach und da hiervon größere Schülerpopulationen betroffen sind, stellt die Suche nach weniger abstrakten und damit verständlicheren Alternativen eine dringende Aufgabe dar.

Das vorliegende Buch verweist auf eine solche Alternative. Es zeigt auf, daß es im Hinblick auf Oberflächenladungen keinen entscheidenden Unterschied gibt zwischen einer elektrostatischen Anordnung und einem stationären Stromfluß. Es verweist auf curriculare Neuentwicklungen zum Spannungsbegriff und gibt einen umfassenden Überblick über die wissenschaftlichen Veröffentlichungen, die seit den Arbeiten von Weber und Kirchhoff erschienen sind.

Warum sollte man also bei der Behandlung des Begriffs "elektrischer Strom" auf das Driften von Elektronen verweisen, beim Begriff "elektrische Spannung" aber nicht auf die Existenz driftender Oberflächenladungen?

Sicherlich wird es das Ziel des Unterrichts sein, den Spannungs- und Potentialbegriff auf der Ebene der Ernergie quantitativ zu behandeln. Eine qualitative und anschauliche Vorstufe auszulassen ist jedoch didaktisch nicht vertretbar, es sei denn, man hat von der Existenz von driftender Oberflächenladungen keine Kenntnis.

Es gibt curriculare Neuentwicklungen zur Elektrizitätslehre, in denen die Inhalte dieses Buches ausführlich zur Sprache kommen. Vergleiche mit traditionellen Kursen hinsichtlich Lernerfolg und Lernmotivation sollten durchgeführt werden, um letzte Zweifel an der Notwendigkeit einer eigenen curricularen Neuentwicklung zu beheben.

Auf dem Weg zu einem tieferen Verständnis elektrischer Phänomene, insbesondere der Bewegung von Elektronen in Stromkreisen liefert dieses Buch einen entscheidenden Beitrag.

Part I

Introduction

The goal of this book is to analyze the force between a point charge and a resistive wire carrying a steady current, when they are at rest relative to one another and the charge is external to the circuit. Analogously, we consider the potential and electric field inside and outside resistive conductors carrying steady currents. We also want to discuss the distribution of charges along the surface of the conductors which generate this field. This is an important subject for understanding the flow of currents along conductors. Unfortunately, it has been neglected by most authors writing about electromagnetism. Our aim is to present the solutions to the main simple cases which can be solved analytically in order to show the most important properties of this phenomenon.

It is written for undergraduate and graduate students in the following courses: physics, electrical engineering, mathematics, history and philosophy of science. We hope that it will be utilized as a complementary text in courses on electromagnetism, electrical circuits, mathematical methods of physics, and history and philosophy of science. Our intention is to help in the training of critical thinking in students and to deepen their knowledge of this fundamental area of science.

We begin by showing that many important authors held incorrect points of view regarding steady currents, not only in the past but also in recent years. We then discuss many experiments proving the existence of a force between a resistive conductor carrying a steady current and an external charge at rest relative to the conductor. This first topic shows that classical electrodynamics is a lively subject in which there is still much to be discovered. The readers can also enhance their critical reasoning in respect to the subject matter.

Another goal is to show that electrostatics and steady currents are intrinsically connected. The electric fields inside and outside resistive conductors carrying steady currents are due to distributions of charges along their surfaces, maintained by the batteries. This unifies the textbook treatments of the subjects of electrostatics and steady currents, contrary to what we find nowadays in most works on these topics.

We begin dealing with pure electrostatics, namely, the force between a conductor and an external point charge at rest relative to it. That is, we deal with electrostatic induction, image charges and related subjects. In particular we calculate in detail the force between a long cylindrical conductor and an external point charge at rest relative to the conductor.

We then move to the main subject of the book. We consider the force between a resistive wire carrying a steady current and a point charge at rest relative to the wire, outside the wire. In particular, we deal with the component of this force which is proportional to the voltage of the battery connected to the wire (we discuss the voltage or electromotive force of a battery, together with its distinction from the concept of potential difference, in Section 5.3). We embark on this analysis by first considering straight conductors of arbitrary cross-section in general and a general theorem on their surface charges. Next we deal with a long straight conductor of circular cross-section. Then we treat a coaxial cable and a transmission line (twin lead). We subsequently deal with conducting planes and a straight strip of finite width.

In the third part we consider cases in which the closed current follows curved trajectories through resistive conductors. Once more we are interested in the force between this conductor and an external point charge at rest relative to it. Initially we deal with a long cylindrical shell with azimuthal current. Then we consider the current flowing in the azimuthal direction along a resistive spherical shell. And finally we treat the case of a toroidal conductor with steady azimuthal current. Although much more complicated than the previous cases, this last situation is extremely important, as it can model a circuit bounded in a finite volume of space carrying a closed steady current, like a resistive ring.

Our intention in including analytical solutions of all these basic cases in a single work is to make it possible to utilize this material in the undergraduate and graduate courses mentioned earlier. Although the mathematical treatments and procedures are more or less the same in all cases, they are presented in detail for conductors of different shapes, so that the chapters can be studied independently from one another. It can then easily be incorporated in standard textbooks dealing with electromagnetism and mathematical methods for scientists. Part of the material presented here was previously discussed in textbooks and research papers. We feel that the reason why it has not yet been incorporated into most textbooks, which actually present false statements related to this topic, is that all these simple cases have never been assembled in a coherent fashion. We hope to overcome this limitation with this book.

At the end of this work we present open questions and future prospects. In an Appendix we discuss an important work by Wilhelm Weber where he presented a calculation of surface charges in resistive conductors carrying a steady current, a remarkable piece of work which has unfortunately been forgotten during all these years. We also discuss Kirchhoff's work on surface charges and the derivation by Weber and Kirchhoff of the telegraphy equation.

A full bibliography is included at the end of the book. In this work we utilize the International System of Units SI. When we define a concept, we utilize the $\equiv$ symbol to denote a definition. We represent the force exerted by body j on i by $\vec{F}_{ji}$. When we say that a body is stationary or moving with velocity $\vec{v}$, we consider the laboratory as the frame of reference, unless stated otherwise. The laboratory is treated here as an approximately inertial frame of reference, for the purpose of experiment. When we say that a "charge" exerts a force, creates an electric field, or is acted upon by an external force, we mean a "charged body," or a "body with the property of being electrically charged." That is, we consider charge as a property of a body, not as a physical entity. We consider the concepts of electric and magnetic fields to be mathematical devices embodying the physical forces between charged bodies, between magnets or between current carrying conductors. That is, it is possible to say that a current-carrying wire generates electric and magnetic fields, as usually expressed by most authors. In this sense an alternative title of this book might be "The electric field outside resistive wires carrying steady currents." But the primary reality for us is the force or interaction between material bodies (generating their relative accelerations relative to inertial frames), and not the abstract field concepts existing in space independent of the presence of a charged test particle which can detect the existence of these fields.

Chapter 1

Main Questions and False Answers

1.1 Simple Questions

Consider a resistive circuit as represented in Figure 1.1.

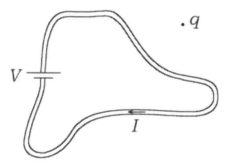

Figure 1.1: A battery supplying a constant voltage V between its terminals generates a steady current I in a uniformly resistive wire. Is there a force between the circuit and an external point charge q at rest relative to the wire? Is any component of this force proportional to the voltage of the battery?

A stationary, homogeneous and isotropic wire of uniform resistivity connected to a battery (which generates a voltage V between its terminals) carries a steady current I. The main questions addressed in this work are the following:

a) Will the resistive wire carrying a steady current exert a force on a stationary charge q located nearby? Will any component of this force depend upon the voltage generated by the battery? This is the most important question discussed in this work.

b) A related question is the following: Will this wire exert any action upon a conductor, or upon neutral dielectrics placed nearby? In particular, will the resistive wire carrying a steady current electrically polarize a neutral conductor placed nearby, attracting the conductor?

We can also rephrase these questions utilizing the concepts of electric and magnetic fields. In this case we can say that the current-carrying wire creates a magnetic field outside itself. This magnetic field will act upon mobile test charges. We can then rephrase our question in terms of an electric field:

c) Does a resistive wire connected to a battery and carrying a steady current produce an external electric field? If so, is this electric field dependent upon the voltage V generated by the battery?

Other related questions:

d) Is the resistive wire carrying a steady current electrically neutral along its surface? If not, how does the density of surface charges vary along the length of the wire? That is, how does it change as a function of the distance along the wire from one of the terminals of the battery? Is this density of surface charges a function of the voltage of the battery?

e) Does the wire carrying a steady current have a net distribution of charges inside it? That is, is it electrically neutral at all internal points? If it is not neutral, does this volume density of charges depend upon the voltage of the battery? Will this volume density of charges vary along the length of the wire, *i.e.*, as a function of the distance along the wire from one of the terminals of the battery?

f) Where are the charges which produce the internal electric field in a current-carrying wire located? This electric field is essentially parallel to the wire at each point, following the shape and curvature of the wire, according to Ohm's law. But where are the charges that create it? Are they all inside the battery (or along the surface of the battery)?

These are the main questions discussed in this work[1] [1].

One force which will be present regardless of the value of the current is that due to the electrostatically induced charges in the wire. That is, the external point particle q induces a distribution of charges along the surface of the conducting wire, and the net result will be an electrostatic attraction between the wire and q. Most authors know about this fact, although the majority forget to mention it. Moreover, they neither consider it in detail nor give the order of magnitude of this force of attraction.

Is there another force between the wire and the stationary charge? In particular, is there a force between the stationary charge and the resistive current carrying wire that depends upon the voltage of the battery connected to the wire? Many physicists believe the answer to this question is "no," and this opinion has been held for a long time. There are three main reasons for this belief. We analyze each one of them here. The quotations presented herein are not intended to be complete, nor as criticism of any specific author, but only to

---

[1] All papers by Assis can be found in PDF format at: http://www.ifi.unicamp.br/~assis/

indicate how widespread false beliefs about basic electromagnetism really are.

1.2 Charge Neutrality of the Resistive Wire

The first idea relates to the supposition that a stationary resistive wire carrying a steady current is essentially neutral in all its interior points and along its entire surface. This leads to the conclusion that a resistive wire carrying a steady current generates only a magnetic field outside it. Many scientists have held this belief, for more than a century. Clausius (1822-1888), for instance, based all his electrodynamics on this supposition. In 1877 he wrote ([2] and [3, page 589]): "We accept as criterion the experimental result that a closed constant current in a stationary conductor exerts no force on stationary electricity." Although he stated that this is an experimental result, he did not cite any experiments that sought to find this force. As we will see, he based his electrodynamics on an untenable principle, as a force between a stationary wire carrying a steady current and an external stationary charge does exist. This force has been shown to exist experimentally, as we discuss below. We confirm the existence of this force with calculations.

Recently the name "Clausius postulate" has been attached by some authors to the following statements: "Any current element of a closed current in a stationary conductor is electrically neutral" [4]; "For a stationary circuit the charge density ρ is zero" [5]; "$\Phi = 0$," namely, that the potential generated by a closed circuit carrying a steady current is null at all external points [6, 7].

We even find statements like this in fairly recent electromagnetic textbooks. As we will see, the electric field inside and outside a resistive wire carrying a steady current is due to surface charges distributed along the wire. On the other hand, Reitz, Milford and Christy, for instance, seem to say that no steady surface charges can exist in resistive wires [8, pp. 168-169]: "Consider a conducting specimen obeying Ohm's law, in the shape of a straight wire of uniform cross-section with a constant potential difference, $\triangle\varphi$, maintained between its ends. The wire is assumed to be homogeneous and characterized by the constant conductivity g. Under these conditions an electric field will exist in the wire, the field being related to $\triangle\varphi$ by the relation $\triangle\varphi = \int \vec{E} \cdot d\vec{\ell}$. It is evident that there can be no steady-state component of electric field at right angles to the axis of the wire, since by Eq. $\vec{J} = g\vec{E}$ this would produce a continual charging of the wire's surface. Thus, the electric field is purely longitudinal." Although Russell criticized this statement as it appeared in second edition of the book (1967) [9], the third and fourth editions were not changed significantly on this point. Here we show that there is a steady surface charge in this conductor, and that there is a steady-state component of electric field at right angles to the axis of the wire, contrary to their statement.

In Jackson's book we find the following statement ([10, exercise 14.12, page 503] [11, exercise 14.13, page 697]): "As an idealization of steady-state currents flowing in a circuit, consider a system of N identical charges q moving with constant *speed* v (but subject to accelerations) in an arbitrary closed path. Successive charges are separated by a constant small interval $\triangle$. Starting with the Liénard-Wiechert fields for each particle, and making no assumptions concerning the speed v relative to the velocity of light show that, in the limit $N \to \infty$, $q \to 0$, and $\triangle \to 0$, but $Nq = $ constant and $q/\triangle = $ constant, no radiation is emitted by the system and the electric and magnetic fields of the system are the usual static values. (Note that for a real circuit the stationary positive ions in the conductors will produce an electric field which just cancels that due to the moving charges.)"

Here Jackson refers to the second order electric field and the lack of radiation produced by all the electrons in a current carrying resistive wire, even though the electrons are accelerated. However, a casual reader of this statement, specially the sentence in parenthesis, will conclude that Clausius was right. However, we will see here that there is a net nonzero electric field outside a stationary resistive wire carrying a steady current. Despite the wording of this exercise, it must be stressed that Jackson is one of the few modern authors who is aware of the electric field outside wires carrying steady currents, as can be seen in his important work of 1996 [12]. In the third edition of this book the sentence between parenthesis has been changed to [13, exercise 14.24, pages 705-706]: "(Note that for a real circuit the stationary positive ions in the conductors neutralize the bulk charge density of the moving charges.)" In this form the sentence does not explicitly mention whether or not an external electric field exists. But even the statement of charge neutrality inside a wire carrying a steady current is subject to debate. See Section 6.4.

Edwards said the following in the first paragraph of his 1974 paper on the measurement of a second order electric field [14]: "For over a century it has been almost axiomatic in electromagnetism that the electric field produced by a current in a stationary conductor forming a closed circuit is exactly zero. To be sure the first order field, dependent upon the current I, is experimentally and theoretically zero, but several early electromagnetic theories, including Weber's, Riemann's and Ritz', predict a second order effect dependent upon I^2 or v^2/c^2 where $\mathbf{v}$ is the charge drifting velocity." We will see here that there is an electric field outside a resistive wire carrying a steady current. This external electric field is proportional to the voltage of the battery, or to the potential difference acting along the wire.

Edwards, Kenyon and Lemon had the following to say about first order terms, *i.e.*, to forces proportional to the current due to a resistive wire carrying a steady current, or forces proportional to v_d/c, where v_d is the drifting velocity of the moving charges in the wire and c is the light velocity [15]: "It has long been known that the zero- and first- order forces on a charged object near a charge- neutral, current-carrying conductor at rest in the laboratory are zero in magnitude." The experiments discussed below and the calculations presented in this book show that a normal resistive wire carrying a steady current cannot be electrically neutral at all points. Moreover, it will generate a zeroth order force upon a charged body placed in proximity. It will also generate a force proportional to the voltage or electromotive force of the battery connected to the wire. This force will act upon any charged body brought near the wire. It will also polarize any neutral conductor that is brought near the wire.

A similar statement can be found in Griffiths's book [16, p. 273]: "Within a material of uniform conductivity, $\nabla \cdot \mathbf{E} = (\nabla \cdot \mathbf{J})/\sigma = 0$ for steady currents (equation $\nabla \cdot \mathbf{J} = 0$), and therefore the charge density is zero. Any unbalanced charge resides on the *surface*." We will show that there is also an unbalanced charge in the interior of a resistive wire carrying a steady current.

And similarly [16, p. 196] (our emphasis in boldface): "Two wires hang from the ceiling, a few inches apart. When I turn on a current, so that it passes up one wire and back down the other, the wires jump apart – they plainly repel one another. How do you explain this? Well, you might suppose that the battery (or whatever drives the current) is actually charging up the wire, so naturally the different sections repel. But this "explanation" is incorrect. **I could hold up a test charge near these wires and there would be no force on it, indicating that the wires are in fact electrically neutral. (It's true that electrons are flowing down the line - that's what a current *is* – but there are still just as many plus as minus charges on any given segment.)** Moreover, I could hook up my demonstration so as to make the current flow up *both* wires; in this case the wires are found to *attract!*" Here we show that the statement in boldface is wrong.

Despite these statements it should be mentioned that Griffiths is aware of the surface charges in resistive conductors with steady currents and the related electric field outside the wires [16, pp. 279 and 336-337].

A similar statement is made by Coombes and Laue [17]: "For a steady current in a homogeneous conductor, the charge density ρ is zero inside the conductor."

Lorrain, Corson and Lorrain, meanwhile, state that [18, p. 287]: "A wire that is stationary in reference frame S carries a current density J. The net volume charge density in S is zero: $\rho = \rho_p + \rho_n = 0$." Here ρ_p and ρ_n refer to the positive and negative volume charge densities, respectively.

Although aware of the distribution of charges along the surface of resistive conductors carrying steady currents and the corresponding external electric field, Seely also believed that the internal density of charges is zero [19, p. 149]: "Note that the net charge in any element of volume inside a conductor must be zero in either the static case or the electron-flow case. That is, the net charge per unit volume when the electrons and the ions of the metal lattice are considered just balance. Otherwise, an unstable component of an electric field will be developed. Hence, all net electric charge in a conductor resides on the surface of the conducting material. It is the function of the generator to pile up electrons on one end of the conductor and to remove them from the other end. The internal field is thus produced by a density gradient of the surface charges."

Like Seely, Popovic was also aware of the surface charges and external electric field of resistive wires with steady currents, and he even presents a qualitative drawing depicting them [20, pp. 201-202]. But in the next section he "proves" that the volume density of free charges ρ goes to zero at all internal points of a homogeneous conductor [20, p. 206]: "$\rho = 0$ (at all points of a homogeneous conductor). This is a very important conclusion. Accumulations of electric charges creating the electric field that maintains a steady current in homogeneous conductors cannot be inside the conductors. Charges can reside only on the boundary surfaces of two different conductors, or of a conductor and an insulator."

The flaw in all these statements is that the authors have forgotten or neglected the azimuthal magnetic field inside conductors with steady currents which is created by the longitudinal current. The magnetic force due to this field acting upon the conduction electrons will lead to an accumulation of negative charges along the axis of the conductors, until an electric field is created orthogonal to the conductor axis. This will exert an electric force on the mobile electrons, balancing the magnetic force. As a consequence, a steady current conductor must have a net negative volume density of charges in its interior, as we will discuss quantitatively in Section 6.4.

Despite this shortcoming, Popovic's important work is one of the few textbooks that calls attention to the external electric field of current carrying resistive wires, and that even presents a qualitative drawing of this field in a generic circuit.

One of us (AKTA) also assumed, in previous publications, that a resistive wire carrying a steady current was essentially neutral at all points. On the topic of positive q_{i+} and negative q_{i-} charges of a current element i, we wrote [21]: "In these expressions we assumed $q_{i-} = -q_{i+}$ because we are considering only neutral current elements." The same assumption was made a year later [22]: "We suppose this current distribution to have a zero net charge $q_{2-} = -q_{2+}$." In 1994 we wrote [23, p. 85]: "To perform this summation we suppose that the current elements are electrically neutral, namely $dq_{j-} = -dq_{j+}$, $dq_{i-} = -dq_{i+}$. This was the situation in Ampère's experiments (neutral currents in metallic conductors), and happens in most practical situations (currents in wires, in gaseous plasmas, in conducting liquid solutions, *etc.*)" And similarly, in a Section entitled "Electric Field Due to a Stationary, Neutral and Constant Current" we wrote [23, p. 161]: "In this wire we have a stationary current I_2 which is constant in time and electrically neutral." Here we show in detail that these statements are not valid for normal resistive wires carrying steady currents. When we wrote these statements we were not completely aware of the external electric field proportional to the voltage of the battery, which is the main subject of this book, nor of its related surface charges. We were following most other textbook authors in assuming resistive wires carrying steady currents to be essentially neutral at all points. We were concerned only with the second order electric field, a subject which we also discuss in this work. It was around 1992 that we began to be aware of the surface charges in resistive wires and the corresponding external electric field proportional to the voltage of the battery, due to a study of Kirchhoff's works from 1849 to 1857 [24, 25, 26]. All three of these important papers by Kirchhoff exist in English translation [27, 28, 29]. We tried to understand, repeat and extend Kirchhoff's derivation of the telegraphy equation based on Weber's electrodynamics. We suceeded in 1996, and the result of our labours was published in 2000 [30] and 2005 [31]. Simultaneously for several years we sought a solution for the potential outside a straight cylindrical resistive wire carrying a steady current, until we found the solution in 1997, presented here in Chapter 6. In the same years we discovered Jefimenko's book and papers with his experiments, and also many papers by other authors cited earlier. In the following Chapters we show how much have we learned from important recent authors who studied surface charges and related topics in specific configurations. We quote them in the appropriate sections. Our first paper on this subject was published in 1999 [1]. Since then we have published other works dealing with several other spatial configurations. We only became aware of Weber's 1852 work [32] dealing with related subjects in 2001-2002 during our research in Germany quoted in the Acknowledgments. In the period from 2004 to 2006 we had the opportunity to study Weber's work in greater detail, and we present a discussion of it in the first Appendix of this book.

Our hope in publishing this book is that others will not need to follow this tortuous path of discovery. In the bibliography at the end of the book, we have collected many important references by recent authors who have dealt with the subject of this book. We hope that others can draw on these works to achieve new results in a more efficient manner.

1.3 Magnetism as a Relativistic Effect

The second idea leading to the conclusion that a normal resistive current-carrying wire generates no electric field outside it arises from the supposition that magnetism is a relativistic effect. A typical statement of this position can be found in *Feynman's Lectures on Physics*, specifically in Section 13-6 (The relativity of magnetic and electric fields [33, p. 13-7]) (our emphasis in boldface): "We return to our atomic description of a wire carrying a current. **In a normal conductor, like copper,** the electric currents come from the motion of some of the negative electrons - called the conduction electrons - while the positive nuclear charges and the remainder of the electrons stay fixed in the body of the material. We let the density of the conduction electrons be ρ_- and their velocity in S be $\mathbf{v}$. The density of the charges at rest in S is ρ_+, which must be equal to the negative of ρ_-, since we are considering an uncharged wire. **There is thus no electric field outside the wire**, and the force on the moving particle is just $\mathbf{F} = q\mathbf{v_o} \times \mathbf{B}$." The statement that there is no electric field outside a resistive wire (like copper) carrying a constant current is certainly false. One of the main goals of this book is to calculate this electric field and compare the theoretical calculations with the experimental results presented below.

In Purcell's *Electricity and Magnetism* we find the same ideas [34]. In Section 5.9 of this book, which treats magnetism as a relativistic phenomenon, he models a current-carrying wire by two strings of charges, positive and negative, moving relative to one another. He then considers two current carrying metallic wires at rest in the frame of the laboratory, writing (p. 178): "In a metal, however, only the positive charges remain fixed in the crystal lattice. Two such wires carrying currents in opposite directions are seen in the lab frame in Fig. 5.23a. The wires being neutral, there is no electric force from the opposite wire on the positive ions which are stationary in the lab frame." That is, he believes that a resistive wire carrying a steady current generates no external electric field. For this reason he believes that this wire will not act upon an external test charge at rest relative to the wire. This is simply false. A normal resistive stationary metallic wire carrying a steady current cannot be neutral at all points. It must have a distribution of surface charges which will produce the electric field driving the current inside it, and which will also exert net forces upon the stationary charges of the other wire.

Other books dealing with relativity present similar statements connected with Lorentz's transformations between electric and magnetic fields, about magnetism as a relativistic effect, about a normal resistive wire carrying a steady current being electrically neutral, *etc*. For this reason we will not quote them here. The examples of Feynman, Leighton, Sands and Purcell illustrate the problems of these points of view.

It is important to recall here that Jackson [11, Section 12.2, pp. 578-581] and Jefimenko [35] have shown that it is impossible to derive magnetic fields from Coulomb's law and the kinematics of special relativity without additional assumptions.

1.4 Weber's Electrodynamics

The third kind of idea related to this widespread belief is connected with the electrodynamics developed by Wilhelm Eduard Weber (1804-1891), in particular his force law of 1846.

Weber's complete works were published in 6 volumes between 1892 and 1894 [36, 37, 38, 39, 40, 41]. Only a few of his papers and letters have been translated into English [42, 43, 44, 45, 46, 47, 48, 49, 50, 51, 52, 53, 54, 55].

The best biographies of Weber are those of Wiederkehr [56, 57, 58]. Some other important biographies and/or discussions of his works can be found in several important publications and in the references quoted in these works [3, 59, 60, 61, 62, 63, 64, 65, 66, 67, 68, 69, 70, 71, 72, 73, 74, 75, 76, 77].

Modern applications, discussions and developments of Weber's law applied to electrodynamics and gravitation can be found in several recent publications [23, 78, 79, 80, 81, 82, 83, 84, 85, 86, 87, 88, 89, 90, 91, 92, 93, 94, 95, 96, 97, 98, 99, 100, 101, 102, 103, 104, 105, 106, 107, 108, 109, 110, 111, 112, 113, 114, 115, 116, 117, 118, 119, 120, 121, 122, 123, 124, 125, 126, 127, 128, 129, 130, 131, 132, 133]. Several other works and authors are quoted in these books and papers.

Weber's force is a generalization of Coulomb's law, including terms which depend on the relative velocity and relative acceleration between the interacting charges. Charges q_1 and q_2 located at $\vec{r}_1$ and $\vec{r}_2$ move with velocities $\vec{v}_1$ and $\vec{v}_2$ and accelerations $\vec{a}_1$ and $\vec{a}_2$, respectively, relative to a frame of reference O. According to Weber's law of 1846 the force exerted by q_2 on q_1, $\vec{F}_{21}$, is given by (in the international system of units and with vectorial notation):

$$\vec{F}_{21} = \frac{q_1 q_2}{4\pi\varepsilon_0} \frac{\hat{r}_{12}}{r_{12}^2} \left(1 - \frac{\dot{r}_{12}^2}{2c^2} + \frac{r_{12}\ddot{r}_{12}}{c^2}\right) = -\vec{F}_{12} \ . \tag{1.1}$$

Here $\varepsilon_0 = 8.85 \times 10^{-12}$ C^{2}N^{-1}m^{-2} is called the permittivity of free space, $r_{12} \equiv |\vec{r}_1 - \vec{r}_2|$ is the distance between the charges, $\hat{r}_{12} \equiv (\vec{r}_1 - \vec{r}_2)/r_{12}$ is the unit vector pointing from q_2 to q_1, $\dot{r}_{12} \equiv dr_{12}/dt = \hat{r}_{12} \cdot (\vec{v}_1 - \vec{v}_2)$ is the relative radial velocity between the charges, $\ddot{r}_{12} \equiv d\dot{r}_{12}/dt = d^2 r_{12}/dt^2 = [(\vec{v}_1 - \vec{v}_2) \cdot (\vec{v}_1 - \vec{v}_2) - (\hat{r}_{12} \cdot (\vec{v}_1 - \vec{v}_2))^2 + (\vec{r}_1 - \vec{r}_2) \cdot (\vec{a}_1 - \vec{a}_2)]/r_{12}$ is the relative radial acceleration between the charges and $c = 3 \times 10^8$ m/s is the ratio between electromagnetic and electrostatic units of charge. This constant was introduced by Weber in 1846 and its value was first determined experimentally by Weber and Kohlrasch in 1854-55 [62, 134, 135, 136]. One of their papers has been translated into English [55]. In the works quoted above there are detailed discussions of this fundamental experiment and its meaning.

The first point to be mentioned here is that the main subject of this book can be derived from Weber's law. As a matter of fact, we are mainly concerned with the force between a resistive wire carrying a steady current and an external point charge at rest relative to the wire. To this end we employ essentially Coulomb's force between point charges or, analogously, Gauss's law and Poisson's law. And these three expressions (the laws of Coulomb, Gauss and Poisson) are a special case of Weber's law when there is no motion between the interacting charges (or when we can disregard the small second order components of Weber's force, which are proportional to the square of the drifting velocity of the charges, in comparison with the coulombian component of Weber's force).

With this force Weber succeeded in deriving from a single expression the whole of electrostatics, magnetostatics, Ampère's force between current elements and Faraday's law of induction.

When he presented his fundamental force law in 1846, Weber supposed that electric currents in normal resistive wires are composed of an equal amount of positive and negative charges moving relative to the wire with equal and opposite velocities, the so-called Fechner hypothesis [137, pp. 135 and 145 of the *Werke*]. This model of the electric current had been presented by Fechner in 1845 [138]. Ideas of a double current of positive and negative charges somewhat similar to these had been presented before by Oersted [139, 140] and by Ampère [141, 142]. At that time no one knew about electrons, they had no idea of the value of the drifting velocities of the mobile charges in current-carrying wires, *etc.* Later on it was found that only the negative electrons move in metallic wires carrying steady currents, while the positive ions remain fixed relative to the lattice. Despite this fact, in his own paper of 1846 Weber already considered that Fechner's hypothesis might be generalized considering positive and negative charges moving with different velocities. Specifically, larger particles might flow slower, while smaller particles might move faster [137, p. 204 of the *Werke*]. That is, the particle with a larger inertial mass would move slower inside a current carrying wire than the particle with a smaller inertial mass. In his paper of 1852 which we analyse in the Appendix, Weber even considers the situation in which the positive charges were fixed in the conductor while only the negative charges did move relative to it! He was certainly one of the first to explore this possibility, being much ahead of his time.

Two main criticisms were made against Weber's electrodynamics after it was discovered that Fechner's hypothesis is wrong. The first is related to Ampère's force between current elements and the second is related to the force between a current-carrying wire and an external point charge at rest relative to the wire. We will discuss each one of these criticisms separately.

Ampère's force $d^2\vec{F}_{21}$ exerted by the current element $I_2 d\vec{\ell}_2$ upon the current element $I_1 d\vec{\ell}_1$ (located at $\vec{r}_2$ and $\vec{r}_1$ relative to a frame of reference O) is given by (in the international system of units and with vectorial notation):

$$d^2\vec{F}_{21} = -\frac{\mu_0}{4\pi}\frac{\hat{r}_{12}}{r_{12}^2}\left[2(d\vec{\ell}_1 \cdot d\vec{\ell}_2) - 3(\hat{r}_{12} \cdot d\vec{\ell}_1)(\hat{r}_{12} \cdot d\vec{\ell}_2)\right] = -d^2\vec{F}_{12} \ . \quad (1.2)$$

Many recent publications deal with Ampère's work and his force law [103, 114, 143, 144, 145, 146, 147, 148, 149, 150, 151, 152, 153, 154, 155, 156, 157]. Further papers are quoted in these books and works.

Weber knew Ampère's force and derived it from his force law, assuming Fechner's hypothesis. However, many people believed wrongly that without Fechner's hypothesis it would be impossible to derive Ampère's law from Weber's force. For this reason they criticized Weber's law as experimentally invalidated. But it has been shown recently that even without Fechner's hypothesis it is possible to derive Ampère's law from Weber's force [78, 21, 23, 92, 103, 114]. That is, even supposing that the positive ions are fixed in the lattice and that only the electrons move in current-carrying wires, we derive Ampère's force between current elements beginning with Weber's force between point charges. This overcame the first criticism of Weber's law discussed here.

The second criticism is connected with the main subject of this book, the force between a stationary charge and a current carrying resistive wire. Supposing Fechner's hypothesis, we conclude that there would be no force between a stationary current-carrying wire and an external charge at rest relative to the wire (apart from the force of electrostatic induction), if the wire were neutral in its interior and along its surface. This has been known since Weber's time. Later on people began to doubt the validity of Fechner's hypothesis. It was with the utilization of the Hall effect in the 1880's and with the discovery of the electron in 1897 that the order of magnitude of the drifting velocity of the conduction charges inside metals was determined. The sign of mobile charges was also discovered [63, pp. 289-290] [3, Chapter XI: Weber-Ritz, Section 2: The electronic theory of conduction, pp. 512-518]. As a result, it was found that Fechner's hypothesis was wrong and that only negative charges move relative to the lattice in normal resistive metallic conductors carrying steady currents. Supposing (a) that only one kind of charge (positive or negative) moves in a current-carrying wire, (b) Weber's force, and (c) that the wire is neutral in its interior and along its surface, people concluded that there would be a net force between this stationary current-carrying wire and a stationary charge nearby.

This force is proportional to v_d^2/c^2, where v_d is the drifting velocity of the conduction charges and $c = 3 \times 10^8$ m/s. Based on the erroneous belief that a current carrying resistive wire exerts no force on a stationary charge nearby, unaware even of the larger force between the wire and the charge, which is proportional to the voltage of the battery connected to the wire, many authors condemned Weber's law as experimentally invalidated.

This trend goes back at least to Maxwell's *Treatise on Electricity and Magnetism* (1873). He considered the force between a conducting wire carrying a constant current and another wire which carries no current, both of them at rest in the laboratory. He then wrote [158, Volume 2, Article 848, page 482], our words between square brackets: "Now we know that by charging the second conducting wire as a whole, we can make $e' + e_1'$ [net charge on the wire without current] either positive or negative. Such a charged wire, even without a current, according to this formula [based on Weber's electrodynamics], would act on the first wire carrying a current in which $v^2 e + v_1^2 e_1$ [sum of the positive and negative charges of the current-carrying wire by the square of their drifting velocities] has a value different from zero. Such an action has never been observed." As with Clausius's comment mentioned earlier, Maxwell did not quote any experiments which tried to observe this force and which failed to find the effect. Nor did he calculate the order of magnitude of this effect. This calculation would determine whether it was feasible to try to detect the effect in the laboratory. Maxwell does not seem to have been aware of the surface charge distribution in wires carrying steady currents, a subject which had already been extensively discussed by Weber twenty years before, as we discuss in the first Appendix of this book.

Clausius's work of 1877 (On a deduction of a new fundamental law of electrodynamics) was directed against Weber's electrodynamics [2]. To the best of our knowledge this paper has never been translated into English. What we quote here is our translation. Clausius supposes only one type of mobile charge in a closed stationary current-carrying wire. He integrates Weber's force exerted by this wire on an external stationary charge and shows that it is different from zero (in his integration he does not take into account the surface charges generating the electric field inside the resistive current-carrying wire). He then writes, our words between square brackets: "Then the galvanic current must, like a body charged with an excess of positive or negative charges, cause a modified distribution of electricity in conducting bodies placed in its neighbourhood. We would obtain a similar effect in conducting bodies around a magnet, when we explain the magnetism through molecular electric currents. However these effects have never been observed, despite the various opportunities we have had to observe it. We then accept the previous proposition, which states that these effects do not occur, as an acknowledged certain experimental proposition. Then, as the result of Equation (4) [Weber's force different from zero acting upon a stationary external charge, exerted by a stationary closed current-carrying wire with only one kind of mobile charges] is against this proposition. It follows *that Weber's fundamental law is incompatible with the point of view that in a stationary conductor with galvanic current only the positive electricity is in motion.*" He also mentions that this conclusion was reached independently by Riecke in 1873, which he became aware of only in 1876. It seems that he was also unaware of Maxwell's previous analysis. In the sixth Section of his paper he once again emphasizes his fundamental theorem, namely, "that there is no force upon a stationary charge exerted by a stationary closed conductor carrying a constant galvanic current."

The first two paragraphs of the seventh Section of this paper are also relevant. We quote them here with our words between square brackets:

"To deal with the quantity X_1 we can utilize a similar experimental proposition, namely: *a stationary quantity of electricity exerts no force upon a stationary closed conductor carrying a constant galvanic current.*"

"This proposition needs clarification. When there is accumulation of one kind of electricity at any place, for example positive electricity, then this electricity exerts the effect of electrostatic influence [or electrostatic induction] upon conducting bodies in its neighbourhood, and this effect will also affect the conductor in which there is a galvanic current. The previous proposition says only that beyond this effect there is no other special effect depending upon the current, and therefore dependent upon the current intensity. It should also be remarked that if a closed galvanic current did suffer such an effect, then a magnet would also suffer the effect. However it has always been observed, that stationary electricity acts upon a stationary magnet only in the same way as it acts upon a nonmagnetic piece of metal of the same form and size. For this reason we will accept the previous proposition without further consideration as a firm experimental proposition."

Clausius shows here that he is completely unaware of the electric field outside resistive wires carrying steady currents, which is proportional to the voltage of the battery. This electric field originates from surface charges which are maintained by the electromotive force exerted by the battery. For this reason it is probable that no analogous electric field should exist outside a magnet. Therefore, Clausius's conclusion that if an electric field existed (as we know nowadays it really exists) outside resistive wires carrying steady currents, then, necessarily, it would also exist outside a permanent magnet, also seems incorrect.

In this paper Clausius obtains a new fundamental law of electrodynamics which does not lead to this force exerted by a closed stationary resistive conductor carrying a steady current upon an external charge at rest relative to the conductor, even if only one kind of electricity is in motion in current-carrying wires. His electrodynamics led to this prediction: "The fundamental law formulated by me leads to the result, without the necessity to make the supposition of double current, that a constant stationary closed galvanic current exerts no force on, nor suffers any force from, a stationary charge" [159] [3, page 589].

Clausius's work was not the first to criticize Weber's electrodynamics in this regard; after all, Maxwell had done so before. Despite this fact his work was very influential and is quoted on this point by many authors.

Writing in 1951 Whittaker criticized Weber's electrodynamics along the same lines [63, page 205] (our emphasis in boldface): "The assumption that positive and negative charges move with equal and opposite velocities relative to the matter of the conductor is one to which, for various reasons which will appear later, objection may be taken; but it is an integral part of Weber's theory, and cannot be excised from it. In fact, if this condition were not satisfied, and if the law of force were Weber's, electric currents **would exert** forces on electrostatic charges at rest (...)". Obviously he is here expressing the view that there are no such forces. As a consequence, Weber's electrodynamics must be wrong in Whittaker's view, because we now know that only the negative electrons move in metallic wires. And applying Weber's electrodynamics to this situation (in which a current in a metallic conductor is due to the motion of conduction electrons, while the positive charges of the lattice remain stationary) implies that a conducting wire should exert force on a stationary electric charge nearby. Whittaker was not aware of the experimental fact that *electric currents exert forces on electrostatic charges at rest.* See the experiments discussed below.

To give an example of how this misconception regarding Weber's electrodynamics has survived we present here the only paragraph from Rohrlich's book (1965) where he mentions Weber's theory [160, p. 9]: "Most of the ideas at that time revolved around electricity as some kind of fluid or at least continuous medium. In 1845, however, Gustav T. Fechner suggested that electric currents might be due to *particles* of opposite charge which move with equal speeds in opposite directions in a wire. From this idea Wilhelm Weber (1804 - 1891) developed the first *particle electrodynamics* (1846). It was based on a force law between two particles of charges e_1 and e_2 at a distance r apart,

$$F = \frac{e_1 e_2}{r^2} \left[1 + \frac{r}{c^2} \frac{d^2 r}{dt^2} - \frac{1}{2c^2} \left(\frac{dr}{dt} \right)^2 \right].$$

This force seemed to fit the experiments (Ampère's law, Biot-Savart's law), but ran into theoretical difficulties and eventually had to be discarded when, among other things, the basic assumption of equal speeds in opposite directions was found untenable."

Other examples of this widespread belief: In 1969 Skinner said, relative to Figure 1.2 in which the stationary closed circuit carries a constant current and there is a stationary charge at P [161, page 163]: "According to Weber's force law, the current of Figure 2.39 [our Figure 1.2] would exert a force on an electric charge at rest at the point P. (...) And yet a charge at P does not experience any force." As with Clausius's and Maxwell's generic statements, Skinner did not quote any specific experiment which tried to find this force. Amazingly the caption of his Figure 2.39 states: "A crucial test of Weber's force law." To most readers, sentences like this convey the impression that the experiment had been performed and Weber's law refuted. But the truth is just the opposite! In fact, several experiments discussed in this book show the existence of a force between a stationary charge and a resistive wire carrying a steady current.

Pearson and Kilambi, in a paper discussing the analogies between Weber's electrodynamics and nuclear forces, made the same kind of criticisms in a Section called "Invalidity of Weber's electrodynamics" [162]. They consider a straight wire carrying a constant current. They calculate the force on a stationary

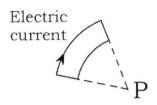

Figure 1.2: There is a stationary charge at P and a steady current flows in the closed circuit.

charge nearby due to this wire with classical electromagnetism and with Weber's law, supposing the wire to be electrically neutral at all points. According to their calculations, classical electromagnetism does not yield any force on the test charge and they interpret this as follows (our emphasis underlined): "The vanishing of the force on the stationary charge q corresponds simply to the fact that a steady current does not give rise to any induced electric field." With Weber's law they find a second order force and interpret this as meaning (our emphasis): "that Weber's electrodynamics give rise to spurious induction effects. This is probably the most obvious defect of the theory, and the only way of avoiding it is to suppose that the positive charges in the wire move with an equal velocity in the opposite direction, which of course they do not." As we will see, the fact is that a steady current gives rise to an external electric field, as shown by the experiments discussed below.

In this work we argue that all of these statements are misleading. That is, we show theoretically the existence of a force upon the stationary external charge exerted by a resistive wire connected to a battery and carrying a steady current when there is no motion between the test charge and the wire. We also compare the theoretical calculations with the experimental results which proved the existence of this force. For this reason these false criticisms of Weber's electrodynamics must be disregarded.

1.5 Electric field of Zeroth Order; Proportional to the Voltage of the Battery; and of Second Order

In this work we discuss the force between a resistive wire carrying a steady current and an external point charge at rest relative to the wire. Both of them are assumed to be at rest relative to the laboratory, which for our purposes can be considered a good inertial frame. We consider three components of this force or electric field.

The wire is a conductor. Let us suppose that it is initially neutral in its interior and along its surface, carrying no current. When we put a charge near it, the free charges in the conductor will rearrange themselves along the surface of the conductor until it acquires a new constant potential at all points. As a result of this redistribution of charges, there will be a net force between the external point charge and the conductor. We will call it a zeroth order force, $\vec{F}_0$. We can describe this situation by saying that the conductor has now produced an induced electrostatic field which will act upon the external charge. This electric field is independent of the current in the conductor, depending only upon the external charge, its distance to the conductor, and the shape of the conductor. That is, this electric field will continue to exist even when a current begins to flow in the conductor, provided the shape of the conductor does not change. We will call it a zeroth-order electric field, $\vec{E}_0$.

We now consider this resistive conductor connected to a battery. In the steady state there is a constant current flowing along the wire. Will there be a force between this wire and the external point charge at rest relative to the wire, with a component of this force depending upon the voltage of the battery? This is the main subject of this book, and the answer is positive. That is, there is a component of this force proportional to the voltage of the battery. We will represent this component of the force by $\vec{F}_1$. It is also possible to say that this wire will generate an external electric field proportional to the voltage of the battery and depending upon the shape of the wire. We will represent this electric field by $\vec{E}_1$.

If there is no test charge outside the wire, the force $\vec{F}_0$ goes to zero, the same happening with $\vec{F}_1$. If it is placed a small conductor, with no net charge, at rest relative to a wire without current, no force $\vec{F}_0$ is observed between them. On the other hand, when this resistive wire is connected to a chemical battery and a constant current is flowing though it, there will be an attractive force between this wire and the small conductor placed at rest nearby. That is, there will be a force $\vec{F}_1$ even when the integrated charge of the small conductor goes to zero. The reason for this force is that the battery will create a redistribution of charges upon the surface of the resistive wire. There will be a gradient of the surface charge density along the wire, with positive charges close to the positive terminal of the battery and negative charges close to the negative terminal of the battery (and with a null charge density at an intermediate point along the wire). This gradient of surface charges will generate not only the internal electric field (which follows the shape of the wire and is essentially parallel to it at each internal point), but also an external electric field. And this external electric field will polarize the small conductor placed in the neighbourhood of the wire. This polarization of the conductor will generate an attractive force between the polarized conductor and the current-carrying wire. With this effect we can distinguish the forces $\vec{F}_0$ and $\vec{F}_1$. This effect has already been observed experimentally, as will be seen in several experiments described in Chapter 3.

Many papers have also appeared in the literature discussing a second order force or a second order electric field, $\vec{F}_2$ or $\vec{E}_2$. As we saw before, usually the people who consider this second order effect are not aware of the zeroth order effects nor of those proportional to the voltage of the battery. This second order force is proportional to the square of the current, or proportional to the square of the drifting velocity of the mobile electrons. Analogously, we can talk of a second order electric field generated by the wire. Sometimes this second order electric field is called motional electric field. It has long been known that the force laws of Clausius and of Lorentz (the ones adopted in classical electromagnetism and presented in almost all textbooks nowadays) do not yield any second order electric field [11, p. 697] [15] [23, Section 6.6]. On the other hand, some theories like those of Gauss, Weber, Riemann and Ritz predict a force of this order of magnitude by taking into account the force of the stationary lattice and mobile conduction electrons acting upon the external stationary test charge [163] [3, Vol. 2, pp. 588-590] [63, pp. 205-206 and 234-236] [162] [15] [23, Section 6.6].

For typical laboratory experiments, as we will show later, this second order force or electric field is much smaller than the force or electric field proportional to the voltage of the battery, which in turn is much smaller than the zeroth order force or electric field. That is, usually we have $|\vec{F}_0| \gg |\vec{F}_1| \gg |\vec{F}_2|$ or $|\vec{E}_0| \gg |\vec{E}_1| \gg |\vec{E}_2|$. In this book we will be concerned essentially with $\vec{E}_0$ and $\vec{E}_1$, discussing only briefly $\vec{E}_2$, due to its extremely small order of magnitude.

Chapter 2

Reasons for the Existence of the External Electric Field

In this Chapter we discuss essentially the electric field proportional to the voltage or to the electromotive force of the battery connected to the circuit. That is, the electric field proportional to the potential difference which is acting along the resistive wire carrying a steady current.

2.1 Bending a Wire

Consider a resistive wire of finite conductivity g connected to a battery and carrying a steady current I, as in the left side of Figure 2.1. The ideal battery generates a constant voltage or electromotive force (emf) V between its terminals.

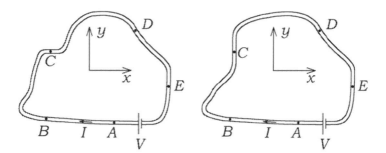

Figure 2.1: The electric field at C points along the x direction in the Figure at left. When the wire is bent as in the right side, the electric field at C now points along the y direction. However, the directions and intensities of the electric fields at A, B, D and E have not changed.

The current density $\vec{J}$ is given by $\vec{J} = (I/A)\hat{u}$, where A is the area of the cross section of the wire and the unit vector $\hat{u}$ points along the direction of the current at every point in the interior of the wire. Ohm's law in differential form states that $\vec{J} = g\vec{E}$, where $\vec{E}$ is the electric field driving the current. By the previous relation we see that $\vec{E}$ will also point along the direction of the wire at each point.

Where are the charges which generate the electric field at each point along the wire located? It might be thought that this electric field is due to the battery (or to the charges located on the battery), but this is not the complete answer. To see that the battery does not generate the electric field at all points along the wire, we can consider Figure 2.1. We know that the electric field driving the constant current will in general follow the shape of the wire. At a specific point C inside the wire the electric field in Figure 2.1 (left circuit) points along the positive x direction. When we bend a portion of the wire, the electric field will follow this bending. In the circuit at the right side of Figure 2.1 it can be seen that at the same point C the electric field now points toward the positive y direction.

If something changes inside the battery when we bend the wire, the electric field at points closer to the battery would also change. However, the electric field changes its path or direction only in the portion which was bent and in the regions close to it, maintaining the previous values and directions in the other points (like the points A, B, D or E in Figure 2.1). As the electric field inside the wire has changed only in the bent portion, something local must have created this change in the direction of the electric field. The shape of the wire has obviously changed. But as the shape or spatial configuration does not create an electric field, the reason must be sought elsewhere. What creates electric fields or the electric forces exerted upon the conduction electrons must be other charges, called here source charges. So there should exist a change in the location of the source charges when we compare the configurations of the left and right sides of Figure 2.1. And this change in the location of the source charges should happen mainly around the bent portion of the wire, but not at the battery. We then arrive at Weber's and Kirchhoff's idea that the electric field inside a wire carrying a constant current is due to free charges spread along the surface of the wire [32, 24, 25, 26]. Kirchhoff's three papers have English translations [27, 28, 29]. In the Appendices we discuss these works in more detail. The role of the battery is to maintain this distribution of free charges along the surface of the wire (constant in time for steady currents, but variable along the length of the wire). There will be a continuous gradient of density of surface charges along the length of the wire, more positive toward the positive terminal of the battery, decreasing in magnitude until it reaches a zero value at an intermediary point, and increasingly negative toward the negative terminal, Figure 2.2.

If there were no battery, the charge density would be zero at all points along the surface of the wire. It is the distribution of these surface charges in space which creates the electric field inside the wire driving the current. When we bend a portion of the wire, the free charges redistribute themselves in space

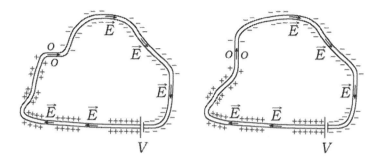

Figure 2.2: Qualitative distribution of charges along the surface of a resistive wire carrying a steady current in two different configurations.

along the surface of the wire, creating the electric field which will follow the new trajectory of the wire. This can be seen qualitatively in the right circuit of Figure 2.2. Supposing the wire to be globally neutral, the integration of the surface charges σ along the whole surface of the wire must always go to zero, although σ is not zero at all points along the surface.

A qualitative representation of the surface charges in a resistive ring carrying a steady current I when connected to a battery generating a voltage V between its terminals is shown on the left side of Figure 2.3. A qualitative representation of the internal and external electric fields due to these surface charges is shown on the right side of Figure 2.3.

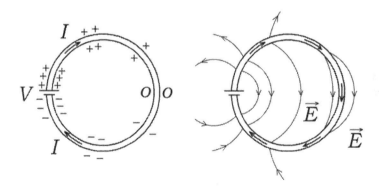

Figure 2.3: Qualitative representation of the surface charges (left) and of the internal and external electric field (right) of a resistive ring carrying a steady current.

A qualitative discussion of this redistribution of surface charges when a wire is bent was given by Parker [164], and by Chabay and Sherwood [165, 166]. An order of magnitude calculation of the charges necessary to bend the electric current I around a corner has been given by Rosser [167].

However, most authors are not aware of these surface charges and the related electric field outside the wire, as we can see from the quotations presented above. Fortunately this subject has been revisited by other authors in some important works discussed in this book.

2.2 Continuity of the Tangential Component of the Electric Field

A second reason for the existence of an electric field outside resistive conductors carrying steady currents is related to the boundary conditions for the electric field $\vec{E}$. As is well known, at an interface between two media 1 and 2 (with $\hat{n}$ being the unit vector normal to the interface at every point) we have that the tangential component of $\vec{E}$ is continuous, $E_{t1} = E_{t2}$ or $\hat{n} \times (\vec{E}_2 - \vec{E}_1) = 0$. On the other hand, the normal component may be discontinuous according to $\hat{n} \cdot (\varepsilon_2 \vec{E}_2 - \varepsilon_1 \vec{E}_1) = \sigma$, where ε_j is the dielectric constant of the medium j and σ is the density of surface charges at the interface. By Ohm's law there must be a longitudinal component of $\vec{E}$ inside a resistive wire, even at its surface. As this component is continuous at an interface, it must also exist in vacuum or in the air outside the conductor, not only close by, but also at measurable distances from the wire.

Many textbooks only consider an electric field outside the current-carrying wire when discussing these boundary conditions. The flux of energy in the electromagnetic field is represented in classical electromagnetism by Poynting's vector $\vec{S} = \vec{E} \times \vec{B}/\mu_0$, where $\vec{B}$ is the magnetic field and $\mu_0 = 4\pi \times 10^{-7}$ H/m is the magnetic permeability of the vacuum. The authors who deal with the electric field outside wires by considering the boundary conditions normally present Poynting's vector pointing radially inwards toward the wire [168, pp. 180-181] [33, p. 27-8]. This goes back to Poynting himself in 1885 [169, 170]. Here is what Poynting wrote in this paper: "In the particular case of a steady current in a wire where the electrical level surfaces cut the wire perpendicularly to the axis, it appears that the energy dissipated in the wire as heat comes in from the surrounding medium, entering perpendicularly to the surface." (...) "In the neighbourhood of a wire containing a current, the electric tubes may in general be taken as parallel to the wire while the magnetic tubes encircle it." In the first paragraph of the section *A straight wire carrying a steady current* he wrote: "Let AB represent a wire in which is a steady current from A to B. The direction of the electric induction in the surrounding field near the wire, if the field be homogeneous, is parallel to AB."

A typical representation found in the textbooks of the fields $\vec{E}$, $\vec{B}$ and $\vec{S}$ in the vicinity of a current carrying cylindrical wire is that of Figure 2.4.

There are two points to make here. In the first place, these drawings and statements suggest that this electric field should exist only close to the wire, while as a matter of fact it exists at all points in space. In the second place,

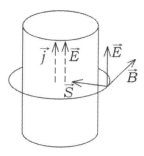

Figure 2.4: Typical representation of the Poynting vector $\vec{S} = \vec{E} \times \vec{B}/\mu_0$ outside a resistive wire carrying a steady current.

they indicate that these authors are not concerned about the surface charges generating the field. It is only at a few locations that $\vec{S}$ will be orthogonal to the wire just outside it, namely, the locations where the surface charge density goes to zero. These locations are an exception and not the rule. In most other locations the density of surface charges will be either positive (closer to the positive terminal of the battery) or negative (closer to the negative terminal of the battery). The rule is that there will be a radial component which may be larger than the longitudinal one, pointing toward the wire or away from it. One of the effects of this radial component is that $\vec{E}$ and $\vec{S}$ will usually be inclined just outside the wire and not orthogonal to it.

These two misleading viewpoints are clearly represented by Feynman, Leighton and Sands's statement in Section 27-5 (Examples of energy flow) of their book [33, p. 27-8], our emphasis in boldface: "As another example, we ask what happens in a piece of resistance wire when it is carrying a current. Since the wire has resistance, there is an electric field along it, driving the current. Because there is a potential drop along the wire, **there is also an electric field just outside the wire, parallel to the surface** (see Fig. 27-5 [our Figure 2.4]). There is, in addition, a magnetic field which goes around the wire because of the current. The **E** and **B** are at right angles; therefore there is a Poynting vector directed radially inward, as shown in the figure. There is a flow of energy into the wire all around. It is, of course, equal to the energy being lost in the wire in the form of heat. So our "crazy" theory says that the electrons are getting their energy to generate heat because of the energy flowing into the wire from the field outside. Intuition would seem to tell us that the electrons get their energy from being pushed along the wire, so the energy should be flowing down (or up) along the wire. But the theory says that the electrons are really being pushed by an electric field, which has come from some charges very far away, and that the electrons get their energy for generating heat from these fields. The energy somehow flows from the distant charges into a wide area of space and then inward to the wire."

As we have seen, the electric field just outside the resistive wire is normally not parallel to the wire. Moreover, the main contribution for the local electric field at a specific point inside a wire is due to the charges along the surface of the wire around this point, contrary to their statement (who believed that it has come "from some charges very far away"). Probably they were thinking here of the charges inside the battery.

Chapter 3

Experiments

In this Chapter we present experiments that prove the existence of the electric field outside resistive wires carrying steady currents. Many experiments along these lines were probably performed in the second half of the XIXth century and in the early part of the XXth century, but they have been forgotten and are not quoted nowadays. Here we present only those which have come to our attention.

We separate these experiments into three classes. The first one is related to the zeroth order electric field (due to electrostatic induction), which exists even when there is no current along the wire. The second class is related directly to the battery and to the current along the wire, being proportional to the emf of the battery or to the potential difference acting along the conductor. The third class is related to the second order electric field, proportional to v_d^2/c^2, where v_d is the drifting velocity of the conduction electrons and c is light velocity in vacuum.

3.1 Zeroth Order Electric Field

We are not aware of any specific experiments designed to measure the force between a point charge and a nearby conductor. We are here considering a conductor which is initially neutral and has no currents flowing through it, until we bring a charge close to it and let both of them at rest relative to one another. After the electrostatic equilibrium is reached, the electrical polarization of the conductor will cause a net force between the conductor and the external charge. This zeroth order force will depend upon the shape of the conductor, upon its distance to the external charge, and upon the value of this charge. We can also express this by saying that a zeroth order electric field will be created depending upon the external test charge, upon its distance to the conductor and upon the shape of the conductor. Many quantitative experiments along these lines were probably performed in the XIXth century. As we are not aware of them, we will not quote any specific experiment here. But we believe these experiments, which were probably made with conductors having many different shapes, would have agreed with the predictions based upon Coulomb's force and upon the properties of conductors, otherwise this would have come to the attention of most scientists long ago. The basic properties of conductors in electrostatic equilibrium which we utilize in this book are: no electric field on the interior, no net density of charges on the interior, any net charge resides on the surface, the potential is constant throughout the interior and the surface of a conductor, and the electric field is perpendicular to the surface immediately outside it. Therefore, we will presume the calculations on this topic to have been confirmed by past observations. In the next section we discuss Sansbury's experiment, which has some qualitative aspects that touch upon this subject.

3.2 Electric Field Proportional to the Voltage of the Battery

We consider here the force between an external test charge and a resistive wire carrying a steady current. We will consider the component of this force which is proportional to the voltage or to the emf of the battery connected to the wire. That is, the component of the electric field proportional to the potential difference acting along the wire. The majority of the experiments deal with voltages of the order of 10^4 V, when the macroscopic effects are more easily seen [171] [166, p. 653].

We present several kinds of experiments. Some map the lines of electric field outside resistive wires carrying steady currents. Others map the equipotential lines outside these conductors. Other experiments directly measure the force between a charge test body and a wire carrying a steady current, when there is no motion between the wire and the test body. Anther experiment measures the charging of an electroscope connected to different points of a circuit carrying a steady current. Yet another experiment describes how to obtain a part of the surface charge in different points of the circuit, showing also how to verify if it is positive or negative and also its magnitude.

Bergmann and Schaefer present some experiments in which they mapped the electric field lines [172, pp. 164-167] [173, pp. 197-199]. They comment that due to the great conductivity of metals it is difficult to utilize metals as conductors in these experiments. Metals cannot sustain a great potential difference between their extremities, so that they produce only a very small external electric field. For this reason they utilize graphite paper strips of high resistivity and apply 20 000 to 40 000 volts between their extremities in order to produce a steady current along the strip. They ground the center of the strip to put it at zero potential, so that the lines of the electric field are symmetrically distributed around it. They then spread semolina in castor oil around the strip, and obtained the result shown in Figure 3.1. The central straight dark line is the paper strip carrying a steady current. The particles of semolina polarize due to the external electric field and align themselves with it, analogous to iron filings mapping a magnetic field.

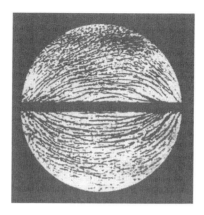

Figure 3.1: Experimental mapping of the external electric field of a straight conductor carrying a steady current.

It should be observed that along the external surface of the conductor there is a longitudinal component of the electric field. This aspect differentiates it from the electric field outside conductors held at a constant potential (in which case the external electric field in steady state is normal to the conductor at every point of its surface), as has been pointed out by Bergmann and Schaefer.

By bending the conductor in a U-form they were able to show the lines of electric field outside a transmission line or twin-lead, Figure 3.2. On the left side the electric field lines are built into the plane of the conductors, while on the right hand side they are built into a plane orthogonal to the conductors.

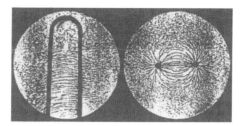

Figure 3.2: Experimental mapping of the external electric field of a transmission line.

Bergmann and Schaefer also discuss the redistribution of charges along the surface of an open circuit connected to a battery when the circuit is closed. Another clear discussion of this situation can be found in the recent book by Chabay and Sherwood [165, Chapter 6].

Experiments similar to those of Bergmann and Schaefer have been performed by Jefimenko [174] [175, pp. 295-312 and 508-511] [176, plates 6 to 9 and pp. 299-319 and 508-511]. He utilized a transparent conducting ink to make a two-dimensional printed circuit on glass plates of 10 inches × 12 inches. In the Figures, 3.3 to 3.7 the gray sections represent the current-carrying conducting strips. The power supply was capable of producing about 10^4 V. He utilized a Du Mont high-voltage power supply type 263-A but mentioned that a small van de Graaff generator might also be employed. After the power supply was turned on, he spread some fine grass seeds (Redtop type) over the plate and conducting system. The seeds lined up in the direction of the electric field over and outside the conductors.

Figure 3.3 depicts Jefimenko's experiment for a straight current-carrying conductor.

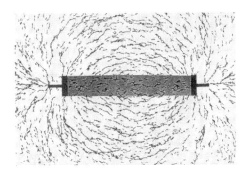

Figure 3.3: Straight current-carrying conductor.

Figure 3.4 depicts his experiment for square-shaped (left) and circular (right) conducting rings.

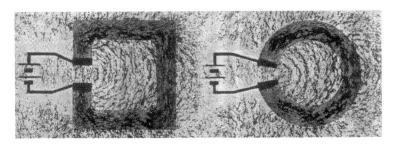

Figure 3.4: Square-shaped (left) and circular (right) conducting rings.

Figure 3.5 depicts his experiment for shorted symmetric (left) and asymmetric (right) transmission lines.

Figure 3.6 depicts his experiment for current-carrying wedges with the two halves connected in parallel (left) and in series (right).

And Figure 3.7 depicts his experiments involving current-carrying rings on the left with two-pole (top) and four-pole (bottom) connections. On the right we have current-carrying discs with two-pole (top) and four-pole (bottom) connections.

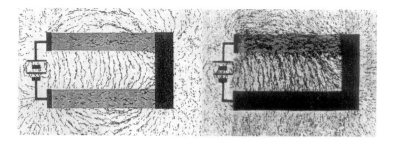

Figure 3.5: Shorted symmetric (left) and asymmetric (right) transmission lines.

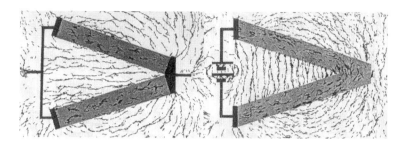

Figure 3.6: Current-carrying wedges with the two halves connected in parallel (left) and in series (right).

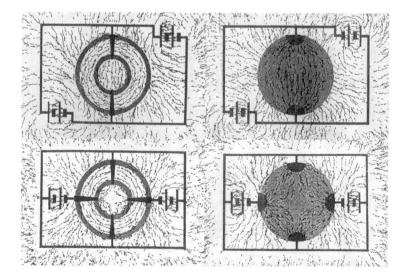

Figure 3.7: On the left are current-carrying rings with two-pole (top) and four-pole (bottom) connections. On the right are current-carrying discs with two-pole (top) and four-pole (bottom) connections.

In a private communication to one of the authors (AKTA), Jefimenko mentioned that he never measured the current in the grass seed experiments. However, he believed that they were of the order of a few microamperes. He mentioned that the patterns of the current-carrying conductors were about 16 or 20 centimeters long.

The experiments of Bergmann, Schaefer and Jefimenko complement one another. After obtaining theoretical formulas for the equipotentials and for the electric field lines, we will compare them with some of these experimental results.

In another type of experiment, Jefimenko, Barnett and Kelly obtained the equipotential lines directly inside and outside conductors with steady currents utilizing an electronic electrometer [177] [176, p. 301]. A radioactive alpha-source was utilized to ionize the air, in order to make it a conductor of electricity, at the point where the field was to be measured. The alpha-source acquired the same potential as the field at that point. The potential was measured (in relation to a reference point chosen at zero potential) with an electronic electrometer connected to the alpha-source. They utilized a hollow rectangular chamber with electrodes for end walls and semi-conducting side walls carrying uniform current. Graphite paper strips were used for the side walls, as in the experiments by Bergmann and Schaefer, and aluminum foil served as electrodes with 80 V applied. The equipotentials were mapped experimentally.

Figure 3.8 depicts the configuration of the problem.

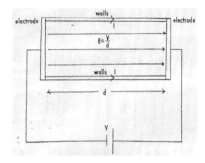

Figure 3.8: Configuration of the system.

Figure 3.9 depicts the equipotential lines measured in one of the experiments.

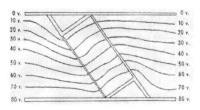

Figure 3.9: Measured equipotentials.

In another experiment, undeveloped photographic film was used in place of the graphite paper, experimentally yielding the equipotential lines inside and outside the conductor carrying steady currents represented in Figures 3.10 and 3.11.

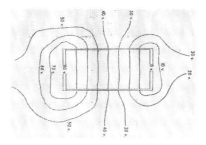

Figure 3.10: Measured equipotentials.

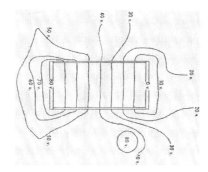

Figure 3.11: Measured equipotentials with 80 Volt disc outside chamber.

In this last experiment, Figure 3.11, they showed that an external charged body has no effect on the field inside the current carrying conductor. The current in the graphite paper was measured to be 5×10^{-2} A, while in the photographic film the current was measured to be only 4×10^{-6} A [177].

A variety of qualitative experiments demonstrating the existence of an external electric field have been performed by Parker [164]. He utilized 5 to 10 kV power supply connected to a high resistance, low-current circuit, which was drawn on a ground-glass or Mylar surface with an IBM scoring pencil. When there was a steady current in the circuit he detected a force on a charged pith ball located nearby, which varied from one end of the circuit to the other. This is a very interesting result, as it shows directly the force between a resistive current carrying circuit and a stationary charge located nearby, the main question we asked in the beginning of this book. Unfortunately Parker did not mention the values of the current, the charge in the pith ball, the distance between the pith ball and circuit, nor the detected force. He also utilized a gold-leaf electroscope with a wire lead to probe quantitatively regions around the current carrying circuit. He could also map the lines of electric field by dusting the glass with plastic or felt fibers while current was flowing. In particular he utilized plastic fibers of approximately 1 mm in length. This mapping is similar to what Bergmann, Schaefer and Jefimenko had done.

There is also an interesting experiment by Sansbury in which he detected a force between a charged metal foil and a current-carrying conductor directly by means of a torsion balance [178]. He placed a neutral 2 cm × 2 cm silver foil which was at the extremity of a torsion balance close to a U-shaped neutral conductor (length 50 cm, separation between the wires 10 cm) without current, Fig. 3.12. When he charged the foil with a charge which he estimated to be approximately 0.5×10^{-9} C (by connecting it to a 3 kV voltage supply), he observed an attraction between the vane and the wire (the charged metal foil moved from a to b in Fig. 3.13). This was probably due to the zeroth order force of electrostatic induction F_0 discussed above, $i.e.$, a force due to image charges induced in the wire by the charged foil nearby. He then passed a steady current of 900 A through the wire by connecting it to a ± 1000 A, 8 V, adjustable, regulated dc current supply. In this case he observed an extra attraction or repulsion between the charged foil and the wire, depending on the sign of the charge in the foil, Fig. 3.13. This extra force was greater than 1.7×10^{-7} N, although he was not able to make precise measurements. This extra force was probably due to the external electric field being discussed here, $i.e.$, to the electric field proportional to the voltage or emf of the battery. Later on we analyze this experiment in more detail in connection with theoretical calculations.

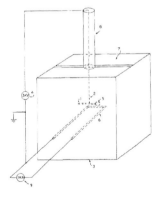

Figure 3.12: Configuration of Sansbury's experiment.

The force between Sansbury's charged metal foil and current-carrying wire seems to be similar to the force between Parker's charged pith ball and current carrying circuit. Bartlett and Maglic considered the force detected by Sansbury an "anomalous electromagnetic effect," as suggested by the title of their paper [179]. They conducted a similar experiment. See Figs. 3.14 and 3.15.

They utilized a rectangular 16-turn coil (4 wide × 4 high), with 30 cm width and 60 cm length. Each turn was made of a 1/8 in. copper tubing (outer diameter of 0.3175 cm). In a private communication with one of the

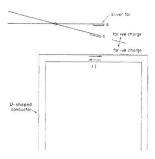

Figure 3.13: In the beginning there is no current in the conductor and the uncharged silver foil remains at a. When the silver foil is charged, it moves from a to b. Then a steady current I is passed through the conductor. In this case there appears an extra force of attraction or repulsion between the charged foil and the U-shaped conductor.

authors (JAH) Bartlett reported that the conductor was water-cooled, with water flowing through the hole in the center of the tubing. Approximately 50% of the cross-sectional area of the tubing was copper, and 50% was water. A current source connected to the coil maintained a steady current of 50 A in each turn. They detected a force upon the charged metal foil with an area of 2.54 cm × 2.54 cm placed at a distance of 3.5 cm from the coil carrying a steady current. This was similar to the effect detected by Sansbury.

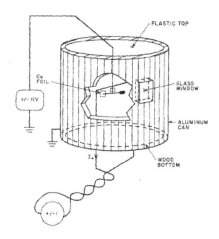

Figure 3.14: Experiment performed by Bartlett and Maglic.

But when the upper half of the current carrying circuit and the test charge were shielded with an aluminum can, the effect disappeared. Their conclusion was that they could not find this "anomalous" interaction, implying that it did

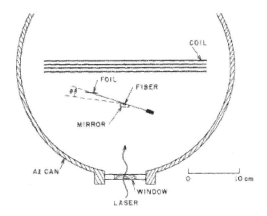

Figure 3.15: Top view of Bartlett and Maglic's experiment.

not exist.

However, they seem to have been unaware of one important point in connection with Faraday cages. They are usually utilized to shield the system under consideration from external influences. But they affect the net force on each internal test charge. For instance, if we have two charges q_1 and q_2 separated by a distance d, the coulombian force between them has a magnitude of $q_1 q_2/4\pi\varepsilon_0 d^2$ and is directed along the line joining them, Figure 3.16.

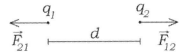

Figure 3.16: Electrostatic force between two charges far from other charges and conductors.

When we surround both of them with a metallic shell, they will induce a distribution of charges along the surface of the shell. As a result of these induced charges, the shell will exert forces on q_1 and on q_2, resulting in general in a net force on each one of them different from the previous value of $q_1 q_2/4\pi\varepsilon_0 d^2$. With a spherical shell the force upon the internal test charges exerted by the induced charges along the wall can be easily calculated by the method of images. Each charge q_j located at a distance a_j from the center of the shell of radius $r_0 > a_j$, with $j = 1$ or 2, will induce charges equivalent to an image charge $q_{ij} = -q_j r_0/a_j$ at a distance $a_{ij} = r_0^2/a_j > r_0$ from the center of the shell, located along the straight line connecting the center of the shell and q_j. The net force on q_1, for instance, will be given by $\vec{F}_{21} + \vec{F}_{i1,1} + \vec{F}_{i2,1}$ instead of simply $\vec{F}_{21}$. Here $\vec{F}_{m,n}$ is the force exerted by the (image) charge m on the charge n. If the straight line connecting the two charges q_1 and q_2 does not pass through the center of the shell, the net force on each one of them will change its magnitude and also its direction in comparison with the previous value without the shell, Figure 3.17.

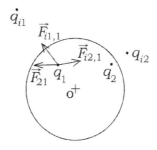

Figure 3.17: Electrostatic forces on q_1 due to q_2, to the image charge q_{i1} and to the image charge q_{i2}.

If there are N internal charges, the net force on q_1 will be given by the sum of the forces due to the other $N-1$ charges on q_1, plus the N forces of the image charges on q_1. If the Faraday cage is not spherical, it will be very difficult to calculate the new force on the test charges. In Bartlett and Maglic's case the Faraday cage was cylindrical with metallic lateral sides and dielectric extremities. This makes it very difficult to estimate the effect of the shield upon the internal charged foil. Moreover, we have not just the two charges q_1 and q_2 as discussed before, but the charged foil and a quantity of charges distributed along the surface of the resistive current-carrying wire. As the wire is made of a conducting material, the induced charges upon the Faraday cage will change the distribution of charges spread along the surface of the wire (compared with the distribution of surface charges without the Faraday cage), such that even this new distribution is not yet known until they can be calculated with Laplace's equation and the appropriate boundary conditions. This enormously complicates the theoretical analysis of the expected net force (exerted upon the test charge when the current-carrying wire and the test charge are shielded). For this reason it seems preferable to perform this kind of experiment without the metallic shield.

The result obtained by Bartlett and Maglic is described as follows [179]: "Averaging the results of the runs with and without the shield we find a signal of 0.3 ± 0.3 mrad. (...) We multiply our measured rotation of 0.3 ± 0.3 mrad by the sensitivity of the fiber (9.1×10^{-5} N/m) to obtain a force of $(0.27 \pm 0.27)10^{-7}$ N." The impression we get from reading the paper is that the force measured without the shield was 0.27×10^{-7} N, one order of magnitude smaller than that observed by Sansbury. On the other hand, we infer that with the shield Bartlett and Maglic measured no force (hence the $\pm$ sign given in the previous quotation). Our opinion is that the shield changes the distribution of charges on the metallic shell. This changes the net force on each of the internal charges, as we showed earlier. Therefore, it is very difficult to compare the two cases (with and without shielding). The ideal situation would be to perform this kind of experiment without any shield. Unfortunately, this was not the procedure adopted by Bartlett and Maglic.

Further discussions of Sansbury's experiment with different approaches can be found in works by several authors [1, 4, 5, 82, 83, 180, 181, 182, 183].

Another kind of experiment was performed by Moreau, Ryan, Beuzenberg and Syme [184]. (See Figure 3.18.)

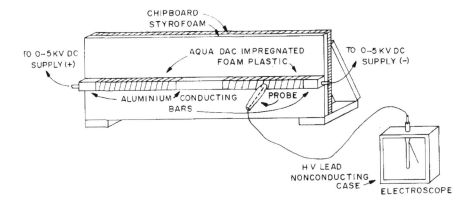

Figure 3.18: Electroscope touching different points of a high voltage resistive circuit carrying a steady current.

They connected a 0- to 5-kV current-limited dc power supply to a series circuit consisting of two resistances of 75 $M\Omega$. The conductors were bare aluminum bars and the resistances were strips of foam plastic impregnated with Aqua Dac. The conductors and resistances had rectangular cross-sections of 12 mm × 9 mm. Each one of the resistances was 50 cm in length. The conducting bar between them was 30 cm in length. The main goal of the experiment was to demonstrate directly that when a steady current flows along a circuit, there is a gradient of surface charges along the conductors and along the resistances, in such a way that this gradient of surface charge density produces an electric field along the direction of the current. To indicate the charge density at several points of the circuit, they connected these points to a high voltage probe. This probe was connected to a gold-leaf electroscope. Something similar to this was done by Parker [164]. The electroscope was placed inside a nonconducting polystyrene case. When the power supply was turned up to 2 kV a microammeter in the circuit measured a current of about 13 μA. One side of the circuit was grounded. At this point the electroscope showed no deflection, indicating zero charge density. As the probe moved away from this point along the resistance strip, the electroscope deflected gradually with distance traveled along the resistance, reaching a deflection of about $55°$ at its end. The deflection remained constant as the probe was moved along the middle conductor, indicating that the charge density was essentially constant (or that it varied very little) along it. The deflection increased again along the second resistance, reaching a final deflection of about 70° at the extremity. This indicated that the charge density was relatively high on the upstream side of the resistors, decreasing along them in the direction of the current. This produced an electric field forcing the conduction charges through the resistances.

Another very didactic experience was performed by Uri Ganiel and collaborators of the Science Education Group at the Weizmann Institute in Israel. This experiment has been cited and reproduced by Chabay and Sherwood [171, 185] [166, Section 18.10, pp x and 652-654]. (See Figure 3.19.)

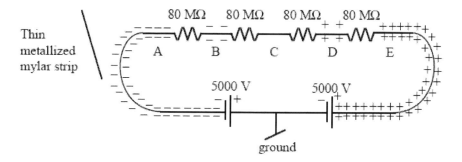

Figure 3.19: Four identical high resistance resistors are connected in series in a closed circuit carrying a steady current. This Figure shows the qualitative distribution of surface charges. The thin metallized mylar strip is attracted by the bare (uninsulated) wires, touches them and is then repelled by them. By testing the charge gained by the metallized strip it is possible to determine the sign of the surface charges at each point along the circuit.

A closed circuit is formed with four identical high resistance resistors connected in series, each with 80 MΩ. They are connected to two high voltage power supplies, of 5 kV each. There is a grounded conductor between these two power supplies. One of the power supplies yields + 5 kV at one extremity of the first resistor, while the other yields - 5 kV at the opposite extremity of the fourth resistor (relative to the ground). In other words, these two power supplies are connected in series. There are bare (uninsulated) wires between the resistors in order to allow the collection of their surface charges by an appropriate probe. This is performed utilizing a flexible, thin metallized mylar strip. When the strip is brought close to the bare wire at the left side of the circuit, near location A, it is observed to be attracted to the wire. It touches the wire and is then repelled by it. It is initially attracted due to polarization of the aluminized plastic strip created by the surface charge on the wire. When it touches the wire, it jumps away, since it is charged by contact with the bare wire and repelled by the surface charge at that location. When the strip is tested it is found to be negatively charged. It is then discharged, and the same procedure is repeated, although this time it is brought close to the bare wire between the first and second resistors, at B. Once more it is found to become negatively charged, but now with a smaller magnitude to the previous case. No effect is observed at C when the strip is first discharged and then brought close to this point. When the procedure is repeated at D, it is found to be positively charged. The same happens at E, but now with a larger magnitude than at D. This experiment shows directly that different points of the wire become charged when there is a constant current flowing through it. This surface charge density changes along the circuit. The largest gradient (variation of the magnitude per unit length) occurs along the resistors. Along the conductors there is a very small variation of the magnitude of the charge density. Chabay and Sherwood point out that only at very high voltages is there enough charge to observe electrostatic repulsion in a mechanical system [166, p. 654]. With a low voltage circuit (like in a flashlight connected to ordinary 1.5 V batteries), any charged body brought near the current-carrying wire will be initially attracted to it, regardless of the sign of the charged body. This will happen both close to the positive and close to the negative terminals of the battery. The reason for this fact is that the zeroth order force will be much greater than the force proportional to the voltage of the battery.

In conclusion we can infer that the experiments of Bergmann, Schaefer, Jefimenko, Barnett, Kelly, Sansbury, Parker, Moreau, Ryan, Beuzenberg, Syme, Ganiel, Chabay and Sherwood prove the existence of the external electric field due to resistive, stationary wires carrying steady currents. They have also shown the existence of surface charges along the conductors and resistors. These are the charges which produce the internal and external electric fields.

In order to have a complete proof, it would be necessary to repeat each of these experiments with different electromotive forces of the batteries and verify if the external electric field is proportional to this voltage. To the best of our knowledge none of these experiments varied the emf in order to show or demonstrate unambiguously linearity of the force intensity (or of the electric field intensity) with applied voltage, unfortunately.

In the following Chapters we compare these experiments with analytical solutions for the external potential and electric field due to resistive wires carrying steady currents. This will give further support to the inference that there is a force proportional to the voltage of the battery between an external stationary point charge and a resistive wire carrying a steady current.

3.3 Second Order Electric Field

All the previous experiments involved the electromotive force of the battery. A completely different set of experiments investigates a second order electric field, i.e., an electric field proportional to v_d^2/c^2.

Checking whether or not the second order electric field exists is much more difficult than studying the electric field discussed previously. The reason for this difficulty is that the order of magnitude of the second order electric field,

E_2, is normally much smaller than the electric field associated with the electromotive force of the battery, E_1, and also much smaller than the zeroth order electric field, E_0. However, if the wire is a superconductor and a steady current is flowing through (with no battery connected to the wire), the external electric field E_1 should go to zero. If we take into account (or neglect) the force due to electrostatic induction (related with image charges) as well, there remains in this case only the second order electric field. This was the approach utilized by Edwards, Kenyon and Lemon in their experiments [14, 15], which are the best known to us to analyze this effect. They utilized type II superconductor (48% niobium and 52% titanium) cores of 2.5 mil radius with currents of the order of 16 A. They found an electric field proportional to I^2, independent of the direction of the current, pointing toward the wire and with an order of magnitude compatible with that predicted by Weber's law. What they actually measured utilizing an electrometer was a potential difference between the circuit and an electrostatic shield around the circuit. They measured potential differences with an order of magnitude of 10 mV.

Bartlett and Ward made a number of different experiments to detect this second order electric field (which they interpreted as due to a possible variation of the electron's charge with its velocity), but failed to find it [186]. They utilized normal resistive conductors, but in their analysis they did not mention the electric field proportional to the voltage of the battery discussed above.

In another experiment, Kenyon and Edwards placed a beam-power radio tube within a Faraday cage [187]. They tried to measure the potential difference between the system and the Faraday cage. They could not find any effect with the order of magnitude of the earlier experiment of Edwards, Kenyon and Lemon.

In any event attention must be called here to the Faraday cage around the system in these experiments, and also in most of those quoted by Bartlett and Ward. As we mentioned previously, the charges induced in the walls of a Faraday cage due to internal charges will exert a net force on any internal test charge. If there are two or more internal charges, the net force in each will be different in two cases: (A) without the Faraday cage (force due only to the other internal charges), and (B) with Faraday cage (force due to the other internal charges and to all induced charges in the walls of the cage). This enormously complicates the analysis of all these experiments, and it is difficult to reach a simple result. Beyond the complication of the Faraday cage, there is also the electric field proportional to the voltage of the battery which must be taken into account before discussing the second order electric field. And this was not done by any of these authors when dealing with resistive conductors. The second order electric field is usually much smaller than the electric field proportional to the voltage of the battery, as we will show later on. For this reason the electric field proportional to the voltage of the battery must be included in the analysis because it can mask the effect which is being sought.

More research is necessary before a final conclusion can be drawn on this second order electric field. A great number of experimental and theoretical works in connection with this subject have been published in the last 25 years

[188, 189, 190, 191, 82, 192, 181, 193, 83, 4, 5, 182, 194, 195, 196, 197, 23, 6, 7, 198].

Chapter 4

Force Due to Electrostatic Induction

4.1 Introduction

The main subject of this book is the force between a stationary wire carrying a steady current and an external charge at rest relative to the wire. In particular, we are interested in the component of this force which is proportional to the voltage or emf of the battery, or to the potential difference acting along the wire.

Before analyzing these cases we consider the force between a point charge and a conductor which has no current flowing through it. We suppose air or vacuum outside the conductor (and also inside hollow ones). We also consider only the equilibrium situation when the point charge and the conductor are at rest relative to one another and also at rest relative to an inertial frame of reference. It is also assumed that there are no other charges or conductors in the vicinity of the system, beyond the ones being considered here. The main material of this Chapter was discussed in 2005 [199].

4.1.1 Point Charge and Infinite Plane

The simplest configuration is that of a point charge q at a distance d from an infinite conducting plane with zero net charge. Let us suppose that the conducting plane is along the plane $z = 0$, while the charge q is located at $(x, y, z) = (0, 0, z)$. The method of images yields in this case an attractive force acting upon q given by

$$\vec{F}_0 = \mp \frac{q^2}{16\pi\varepsilon_0} \frac{\hat{z}}{z^2} \ . \tag{4.1}$$

Here the top sign is valid for $z > 0$, while the bottom sign is valid for $z < 0$.

Expressing this force as $\vec{F}_0 = q\vec{E}_0$ yields a zeroth order electric field given by

$$\vec{E}_0 = \mp \frac{q}{16\pi\varepsilon_0} \frac{\hat{z}}{z^2} \ . \tag{4.2}$$

Even adding a finite charge Q uniformly spread along the infinite plane does not change these two results. That is, $\vec{F}_0$ and $\vec{E}_0$ are independent of Q.

This force is always attractive and diverges to infinity when $d \to 0$.

4.1.2 Point Charge and Spherical Shell

Another simple case to consider is that of a point charge and a conducting spherical shell at rest relative to one another. We consider a spherical shell of radius R centered upon the origin 0 of a coordinate system. There is a net charge Q on the conducting spherical shell, insulated from the earth. The test charge q is located at $\vec{r} = r\hat{r}$ relative to 0. The solution of this problem can also be obtained by the method of images and is found in most textbooks on electromagnetism. When $r > R$ the force upon q is given by

$$\vec{F}_0 = \frac{q}{4\pi\varepsilon_0} \left[Q - \frac{qR^3(2r^2 - R^2)}{r(r^2 - R^2)^2} \right] \frac{\vec{r}}{r^3} \ . \tag{4.3}$$

When $r < R$ the force is independent of Q and is given by

$$\vec{F}_0 = \frac{q^2}{4\pi\varepsilon_0} \frac{R\vec{r}}{(R^2 - r^2)^2} \ . \tag{4.4}$$

These forces diverge to infinity when $r \to R$. When q is inside the shell, it always suffers an electrostatic force toward the closest wall. When q is outside the shell, the force will be attractive not only when $qQ < 0$, but also when $qQ > 0$, provided q is at a very close distance to the shell. A detailed discussion of this fact can be found, for instance, in Maxwell's work [200, Chapter VII: Theory of electrical images, pp. 80-88], in a paper by Melehy [201] and in Jackson's book [11, Section 2.3].

The zeroth order electric field in these cases is given by (with $\vec{F}_0 = q\vec{E}_0$):

$$\vec{E}_0 = \frac{1}{4\pi\varepsilon_0} \left[Q - \frac{qR^3(2r^2 - R^2)}{r(r^2 - R^2)^2} \right] \frac{\vec{r}}{r^3} \ , \quad \text{if } r > R \ . \tag{4.5}$$

$$\vec{E}_0 = \frac{q}{4\pi\varepsilon_0} \frac{R\vec{r}}{(R^2 - r^2)^2} \ , \quad \text{if } r < R \ . \tag{4.6}$$

4.2 Point Charge and Cylindrical Shell

After considering these two simple cases we analyze the main subject of this chapter. The goal is to calculate the electrostatic force between an infinite conducting cylinder of radius a held at zero potential and an external point charge q. To the best of our knowledge this has never been done before. To this end we consider the Green function method [13, Chaps. 1 to 3]. We begin reviewing a known solution of the potential inside a grounded, closed, hollow and finite cylindrical shell with an internal point charge [13, p. 143]. We analyze the limit of an infinite cylinder and explore the force exerted on the point charge. We then perform a similar analysis for the case of an external point charge. We consider in detail the particular situation of a thin wire, i.e., with the point charge many radii away from the axis of the cylinder. These calculations were published in 2005 [199].

4.3 Finite Conducting Cylindrical Shell with Internal Point Charge: Solution of Poisson's Equation

Consider a finite conducting cylindrical shell of radius a and length $\ell \gg a$, with z being its axis of symmetry. (See Fig. 4.1.) With cylindrical coordinates (ρ, φ, z) the center of the shell is supposed to be at $(\rho, z) = (0, \ell/2)$. We also consider a point charge q located at $\vec{r}\,' = (\rho' < a, \varphi', z')$ inside the shell. We wish to calculate the electric potential of the system, the electric field, the surface charge distribution induced by q and the net force between the cylinder and q.

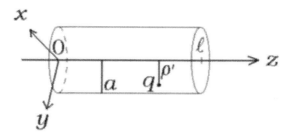

Figure 4.1: Finite conducting cylinder of length ℓ and radius a centered at $(\rho, z) = (0, \ell/2)$, with z being its axis of symmetry. There are conducting covers at $z = 0$ and at $z = \ell$. There is a point charge q located at $(\rho, \varphi, z) = (\rho' < a, \varphi', 0 < z' < \ell)$.

The electrostatic potential ϕ obeys Poisson's equation:

$$\nabla_r^2 \phi = -\frac{\rho}{\varepsilon_0} . \qquad (4.7)$$

By the standard Green function method, the solution of Poisson's equation for this case with Dirichlet boundary conditions (potential specified on a closed surface) is given by:

$$\phi(\vec{r}) = \frac{1}{4\pi\varepsilon_0} \int\int\int_V \rho(\vec{r}'')G(\vec{r},\vec{r}'')dV'' - \frac{1}{4\pi} \oint_S \phi(\vec{r}'')\frac{\partial G}{\partial n''}da'' , \quad (4.8)$$

where V is the volume of the cylindrical shell, S its closed surface and $\partial/\partial n''$ is the normal derivative at the surface S of the shell directed outwards. Here $G(\vec{r},\vec{r}'')$ is a Green function satisfying the equation:

$$\nabla^2_{r''}G(\vec{r},\vec{r}'') = -4\pi\delta(\vec{r}-\vec{r}'') . \quad (4.9)$$

As the surface of the cylinder in electrostatic equilibrium is at a constant potential ϕ_0, we stipulate that $G(\vec{r},\vec{r}'') = 0$ at this surface.

We can expand the Dirac delta function in cylindrical coordinates as given by:

$$\delta(\vec{r}-\vec{r}'') = \delta(\rho-\rho'')\frac{\delta(\varphi-\varphi'')}{\rho}\delta(z-z'') . \quad (4.10)$$

The delta functions for φ and z can be written in terms of orthonormal functions:

$$\delta(z-z'') = \frac{2}{\ell}\left[\sum_{n=1}^{\infty} \sin\frac{n\pi z}{\ell}\sin\frac{n\pi z''}{\ell}\right] , \quad (4.11)$$

$$\delta(\varphi-\varphi'') = \frac{1}{2\pi}\left[\sum_{m=-\infty}^{\infty} e^{im(\varphi-\varphi'')}\right] . \quad (4.12)$$

Notice our particular choice of expansion for z, Eq. (4.11). This choice satisfies the condition $G(\vec{r},\vec{r}'') = 0$ in the covers of the cylindrical shell located at $z=0$ and at $z=\ell$. The Green function can be expanded in a similar fashion:

$$G(\vec{r},\vec{r}'') = \frac{1}{\pi\ell}\left\{\sum_{m=-\infty}^{\infty} e^{im(\varphi-\varphi'')}\left[\sum_{n=1}^{\infty} \sin\frac{n\pi z}{\ell}\sin\frac{n\pi z''}{\ell}g_m(k,\rho,\rho'')\right]\right\} , \quad (4.13)$$

where $k = n\pi/\ell$ and $g_m(k,\rho,\rho'')$ is the radial Green function to be determined. Substituting this expression into Eq. (4.9) and using (4.10) to (4.12) we obtain:

$$\frac{1}{\rho}\frac{d}{d\rho}\left(\rho\frac{dg_m}{d\rho}\right) - \left(k^2 + \frac{m^2}{\rho^2}\right)g_m = -\frac{4\pi}{\rho}\delta(\rho-\rho'') . \quad (4.14)$$

For $\rho \neq \rho''$ the right hand side of Eq. (4.14) is equal to zero. This means that g_m is a linear combination of modified Bessel functions, $I_m(k\rho)$ and $K_m(k\rho)$. Suppose that $\psi_1(k\rho)$ satisfies the boundary conditions for $\rho < \rho''$ and that $\psi_2(k\rho)$ satisfies the boundary conditions for $\rho > \rho''$:

$$\psi_1(k\rho_<) = AI_m(k\rho_<) + BK_m(k\rho_<) , \qquad (4.15)$$
$$\psi_2(k\rho_>) = CI_m(k\rho_>) + DK_m(k\rho_>) . \qquad (4.16)$$

Here A, B, C and D are coefficients to be determined. The symmetry of the Green function in ρ and ρ'' requires that:

$$g_m(k, \rho, \rho'') = \psi_1(k\rho_<)\psi_2(k\rho_>) , \qquad (4.17)$$

where $\rho_>$ and $\rho_<$ are, respectively, the larger and the smaller of ρ and ρ''. The potential must not diverge for $\rho \to 0$, so we must have $B = 0$. The Green function must vanish at $\rho = a$, i.e., $\psi_2(ka) = 0$. This yields $C = -DK_m(ka)/I_m(ka)$. The function g_m can then be written as:

$$g_m(k, \rho, \rho'') = HI_m(k\rho_<)\left[K_m(k\rho_>) - I_m(k\rho_>)\frac{K_m(ka)}{I_m(ka)}\right] . \qquad (4.18)$$

The normalization coefficient $H = AC$ is determined by the discontinuity implied by the delta function in Eq. (4.14):

$$\left.\frac{dg_m}{d\rho}\right|_+ - \left.\frac{dg_m}{d\rho}\right|_- = -\frac{4\pi}{\rho''} = kW[\psi_1, \psi_2] , \qquad (4.19)$$

where the $\pm$ signs means evaluation at $\rho = \rho'' \pm \epsilon$ and taking the limit $\epsilon \to 0$. In the last equality, $W[\psi_1, \psi_2]$ is the Wronskian of ψ_1 and ψ_2. Substituting g_m into Eq. (4.19), and using $W[I_m(k\rho''), K_m(k\rho'')] = -1/(k\rho'')$, we find $H = 4\pi$. The Green function for the problem of a finite conducting cylinder with an internal charge can be finally written as:

$$G(\vec{r}, \vec{r}'') = \frac{4}{\ell}\left\{\sum_{m=-\infty}^{\infty} e^{im(\varphi-\varphi'')}\left\{\sum_{n=1}^{\infty}\sin(kz)\sin(kz'')I_m(k\rho_<)\times\right.\right.$$
$$\left.\left.\times\left[K_m(k\rho_>) - I_m(k\rho_>)\frac{K_m(ka)}{I_m(ka)}\right]\right\}\right\} . \qquad (4.20)$$

4.3.1 Cylindrical Shell Held at Zero Potential

Consider the cylinder to be held at zero potential, namely, $\phi(\vec{r}'') = 0$:

$$\phi(a, \varphi, 0 \leq z \leq \ell) = \phi(\rho \leq a, \varphi, \ell) = \phi(\rho \leq a, \varphi, 0) = 0 . \qquad (4.21)$$

Substituting Eqs. (4.20) and (4.21) into Eq. (4.8) yields the potential inside the cylinder as (with $\rho(\vec{r}'') = q\delta(\vec{r}' - \vec{r}'')$):

$$\phi(\vec{r}, \vec{r}') = \frac{q}{\pi\varepsilon_0\ell}\left\{\sum_{m=-\infty}^{\infty}\left\{\sum_{n=1}^{\infty} e^{im(\varphi-\varphi')}\sin\left(\frac{n\pi z}{\ell}\right)\sin\left(\frac{n\pi z'}{\ell}\right)I_m\left(\frac{n\pi\rho_<}{\ell}\right)\times\right.\right.$$

$$\times \left[K_m\left(\frac{n\pi\rho_>}{\ell}\right) - I_m\left(\frac{n\pi\rho_>}{\ell}\right) \frac{K_m\left(\frac{n\pi a}{\ell}\right)}{I_m\left(\frac{n\pi a}{\ell}\right)} \right] \right\} . \quad (4.22)$$

Here $\rho_>$ ($\rho_<$) is the larger (smaller) of ρ and ρ'.

4.4 Infinite Conducting Cylindrical Shell with Internal Point Charge

We now expand Jackson's solution for the case of a cylindrical shell of infinite length.

The solution for an infinite cylinder differs from the solution of the finite cylinder in that it essentially changes the expansion of the delta function in Eq. (4.11). In the infinite cylinder there is no restriction on the choice of n (or k):

$$\delta(z - z'') = \frac{1}{2\pi} \int_{-\infty}^{\infty} e^{ik(z-z'')} dk = \frac{1}{\pi} \int_0^{\infty} \cos[k(z-z'')] dk . \quad (4.23)$$

The Green function can be written as:

$$G(\vec{r}, \vec{r}'') = \frac{2}{\pi} \left\{ \sum_{m=-\infty}^{\infty} e^{im(\varphi-\varphi'')} \int_0^{\infty} \cos[k(z-z'')] I_m(k\rho_<) \times \right.$$

$$\left. \times \left[K_m(k\rho_>) - I_m(k\rho_>) \frac{K_m(ka)}{I_m(ka)} \right] dk \right\} . \quad (4.24)$$

Note that we can pass from Eq. (4.11) to Eq. (4.23) by transforming the Fourier series into the Fourier transform, that is, by letting $\ell \to \infty$, setting $n\pi/\ell = k$, $dk = \pi/\ell$, $z \to z + \ell/2$, $z'' \to z'' + \ell/2$ and by replacing the infinite sum by the integral over k.

4.4.1 Cylindrical Shell Held at Zero Potential

Consider the cylinder to be held at zero potential, namely, $\phi(a, \varphi, z) = 0$. Substituting Eq. (4.24) into Eq. (4.8), the potential inside the cylinder can be written as (with $\rho(\vec{r}'') = q\delta(\vec{r}' - \vec{r}'')$):

$$\phi(\vec{r}, \vec{r}') = \frac{q}{2\pi^2 \varepsilon_0} \left\{ \sum_{m=-\infty}^{\infty} e^{im(\varphi-\varphi')} \int_0^{\infty} \cos[k(z-z')] I_m(k\rho_<) \times \right.$$

$$\left. \times \left[K_m(k\rho_>) - I_m(k\rho_>) \frac{K_m(ka)}{I_m(ka)} \right] dk \right\} . \quad (4.25)$$

Once more $\rho_>$ ($\rho_<$) is the larger (smaller) of ρ and ρ'.

The zeroth order electric field is given by $\vec{E}_0 = -\nabla\phi$, with components:

$$E_\rho(\rho < \rho') = -\frac{\partial\phi}{\partial\rho} = -\frac{q}{2\pi^2\varepsilon_o}\left\{\sum_{m=-\infty}^{\infty} e^{im(\varphi-\varphi')}\int_0^\infty k\cos[k(z-z')]I_m'(k\rho)\times\right.$$

$$\left.\times\left[K_m(k\rho') - I_m(k\rho')\frac{K_m(ka)}{I_m(ka)}\right]dk\right\}, \tag{4.26}$$

$$E_\rho(\rho > \rho') = -\frac{\partial\phi}{\partial\rho} = -\frac{q}{2\pi^2\varepsilon_0}\left\{\sum_{m=-\infty}^{\infty} e^{im(\varphi-\varphi')}\int_0^\infty k\cos[k(z-z')]I_m(k\rho')\times\right.$$

$$\left.\times\left[K_m'(k\rho) - I_m'(k\rho)\frac{K_m(ka)}{I_m(ka)}\right]dk\right\}, \tag{4.27}$$

$$E_\varphi = -\frac{1}{\rho}\frac{\partial\phi}{\partial\varphi} = \frac{q}{\pi^2\varepsilon_0\rho}\left\{\sum_{m=1}^{\infty} m\sin[m(\varphi-\varphi')]\int_0^\infty \cos[k(z-z')]I_m(k\rho_<)\times\right.$$

$$\left.\times\left[K_m(k\rho_>) - I_m(k\rho_>)\frac{K_m(ka)}{I_m(ka)}\right]dk\right\}, \tag{4.28}$$

$$E_z = -\frac{\partial\phi}{\partial z} = \frac{q}{2\pi^2\varepsilon_0}\left\{\sum_{m=-\infty}^{\infty} e^{im(\varphi-\varphi')}\int_0^\infty k\sin[k(z-z')]I_m(k\rho_<)\times\right.$$

$$\left.\times\left[K_m(k\rho_>) - I_m(k\rho_>)\frac{K_m(ka)}{I_m(ka)}\right]dk\right\}. \tag{4.29}$$

The zeroth order force $\vec{F}_0 = q\vec{E}_0(\vec{r}')$ acting upon the charge q is given by Eq. (4.26) at $\vec{r} = \vec{r}'$ without the first term between brackets (which is the field generated by the charge q itself). There is only a radial component when the cylinder has an infinite length:

$$\vec{F}_0(\vec{r}') = \frac{q^2}{2\pi^2\varepsilon_0}\left\{\sum_{m=-\infty}^{\infty}\int_0^\infty kI_m(k\rho')I_m'(k\rho')\frac{K_m(ka)}{I_m(ka)}dk\right\}\hat{\rho}$$

$$= -\frac{q^2}{4\pi^2\varepsilon_0\rho'^2}\left\{\sum_{m=-\infty}^{\infty}\int_0^\infty I_m^2(x)\frac{d}{dx}\left[x\frac{K_m(xa/\rho')}{I_m(xa/\rho')}\right]dx\right\}\hat{\rho}. \tag{4.30}$$

In the last equation we integrated by parts. We plotted in Fig. 4.2 the zeroth order force of Eq. (4.30), normalized by the constant $F_q \equiv q^2/4\pi\varepsilon_0 a^2$, as a function of ρ'/a. This force goes to zero when $\rho'/a = 0$ and diverges when $\rho' \to a$, as expected.

The surface charges can be calculated using Gauss's law, yielding:

$$\sigma(a,\varphi,z) = \varepsilon_0 E_\rho(a,\varphi,z) =$$

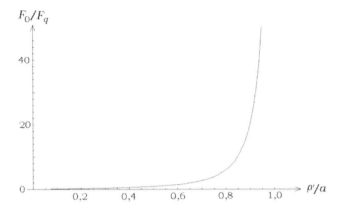

Figure 4.2: Zeroth order force F_0 between an infinite grounded conducting cylinder of radius a and a point charge q at a distance $\rho' < a$ from the z axis (which is also the axis of symmetry of the cylinder), normalized by the constant $F_q \equiv q^2/4\pi\varepsilon_0 a^2$.

$$= -\frac{q}{2\pi^2 a} \left[\sum_{m=-\infty}^{\infty} e^{im(\varphi - \varphi')} \int_0^\infty \cos[k(z-z')] \frac{I_m(k\rho')}{I_m(ka)} dk \right] . \qquad (4.31)$$

The charge per unit length $\lambda(z)$ is given by:

$$\lambda(a, z) = \int_0^{2\pi} \sigma(a, \varphi, z) a\, d\varphi = -\frac{q}{\pi} \int_0^\infty \cos[k(z-z')] \frac{I_0(k\rho')}{I_0(ka)} dk . \qquad (4.32)$$

The total charge induced in the cylinder, supposing $z' = 0$, can be obtained integrating Eq. (4.32) from $z = -\infty$ to ∞. Utilizing:

$$\delta(k) = \frac{1}{2\pi} \int_{-\infty}^\infty \cos(kz) dz = \frac{1}{\pi} \int_0^\infty \cos(kz) dz , \qquad (4.33)$$

this yields:

$$Q = \int_{-\infty}^\infty \lambda(a, z) dz = -q . \qquad (4.34)$$

4.5 Infinite Conducting Cylindrical Shell with External Point Charge

We can now consider a new case, i.e., a conducting cylinder with an external point charge. Suppose that the point charge q is located at $\vec{r}' = (\rho', \varphi', z')$, with $\rho' > a$. Green function can be written analogously in this case as:

$$G(\vec{r}, \vec{r}'') = \frac{1}{2\pi^2} \left[\sum_{m=-\infty}^{\infty} e^{im(\varphi - \varphi'')} \int_0^\infty \cos[k(z-z'')] g_m(k, \rho, \rho'') dk \right] , \qquad (4.35)$$

where g_m can be written as the product $\psi_1'(\rho < \rho'')\psi_2'(\rho > \rho'')$. The functions ψ_1' and ψ_2' satisfy the modified Bessel equation. They can be written as a linear combination of the possible solutions:

$$\psi_1'(k\rho_<) = A' I_m(k\rho_<) + B' K_m(k\rho_<), \qquad (4.36)$$
$$\psi_2'(k\rho_>) = C' I_m(k\rho_>) + D' K_m(k\rho_>). \qquad (4.37)$$

For $\rho \to \infty$ Green function must remain finite. This means that $C' = 0$. Additionally, Green function must be zero at the boundary surface. That is, $G = 0$ at the surface of the cylinder $\rho = a$. This yields:

$$\psi_1'(a) = A' I_m(ka) + B' K_m(ka) = 0 \quad \to \quad B' = -A' \frac{I_m(ka)}{K_m(ka)}. \qquad (4.38)$$

In order to obtain the function g_m we still have to find the constant H':

$$g_m(k, \rho, \rho'') = H' \left[I_m(k\rho_<) - K_m(k\rho_<) \frac{I_m(ka)}{K_m(ka)} \right] K_m(k\rho_>), \qquad (4.39)$$

where $\rho_>$ ($\rho_<$) is the larger (smaller) of ρ and ρ''.
From Eq. (4.19) we have that $H' = 4\pi$:

$$g_m(k, \rho, \rho'') = 4\pi \left[I_m(k\rho_<) - K_m(k\rho_<) \frac{I_m(ka)}{K_m(ka)} \right] K_m(k\rho_>). \qquad (4.40)$$

The Green function is then given by:

$$G(\vec{r}, \vec{r}'') = \frac{2}{\pi} \left\{ \sum_{m=-\infty}^{\infty} e^{im(\varphi - \varphi'')} \int_0^{\infty} \cos[k(z - z'')] \times \right.$$
$$\left. \times \left[I_m(k\rho_<) - K_m(k\rho_<) \frac{I_m(ka)}{K_m(ka)} \right] K_m(k\rho_>) dk \right\}. \qquad (4.41)$$

4.5.1 Cylindrical Shell Held at Zero Potential

Suppose that the surface of the cylinder is held at zero potential, namely:

$$\phi(a, \varphi, z) = 0. \qquad (4.42)$$

Applying Eqs. (4.41) and (4.42) in Eq. (4.8) with $\rho(\vec{r}'') = q\delta(\vec{r}' - \vec{r}'')$ yields:

$$\phi(\vec{r}, \vec{r}') = \frac{q}{2\pi^2 \varepsilon_0} \left\{ \sum_{m=-\infty}^{\infty} e^{im(\varphi - \varphi')} \int_0^{\infty} \cos[k(z - z')] \times \right.$$
$$\left. \times \left[I_m(k\rho_<) - K_m(k\rho_<) \frac{I_m(ka)}{K_m(ka)} \right] K_m(k\rho_>) dk \right\}. \qquad (4.43)$$

Here $\rho_>$ ($\rho_<$) is the larger (smaller) of ρ and ρ'.

Far from the origin, ρ is much larger than ρ', hence we can express Eq. (4.43) in approximate form. The first term that appears between brackets is given by $I_m(k\rho_<)K_m(k\rho_>)$, with $\rho_< = \rho'$ and $\rho_> = \rho$. Note the presence of the term $K_m(k\rho)$, with $\rho \gg \rho'$, which decays rapidly for increasing k. This implies that the main contribution of the integrand is in the region $0 < k < 1/\rho$. Then we can approximate $I_m(k\rho')$ for small arguments, i.e., for $k\rho' \ll 1$, yielding $I_m(k\rho') \approx (k\rho'/2)^m/m!$. From this we can see that the most relevant term is the first one, $m = 0$. The integral of the first term between brackets in Eq. (4.43) is then given by:

$$\phi_1(\rho \gg \rho') \approx \frac{q}{2\pi^2\varepsilon_0} \int_0^\infty \cos[k(z-z')]K_0(k\rho)dk = \frac{q}{4\pi\varepsilon_0\rho} \ . \quad (4.44)$$

To arrive at the last equality we have used the identity given by [202, Prob. 11.5.11]:

$$\frac{2}{\pi}\int_0^\infty \cos(xt)K_0(yt)dt = \frac{1}{\sqrt{x^2+y^2}} \ . \quad (4.45)$$

The second term that appears between brackets in Eq. (4.43) can be treated in a similar way. The main contribution of the integrand is in the region $0 < k < 1/\rho$. Again, the most relevant term is the first one. Accordingly, we approximate the function $K_0(k\rho')$ for small arguments: $K_0(k\rho') \approx -\ln(k\rho')$. The integral of the second term between brackets of Eq. (4.43) is then given by:

$$\phi_2(\rho \gg \rho') \approx -\frac{q}{2\pi^2\varepsilon_0}\int_0^\infty \cos[k(z-z')]\frac{\ln(k\rho')}{\ln(ka)}K_0(k\rho)dk \ . \quad (4.46)$$

From Eq. (4.43) the electric field is given by $\vec{E}_0 = -\nabla\phi$, with components:

$$E_\rho(\rho < \rho') = -\frac{q}{2\pi^2\varepsilon_0}\Bigg\{\sum_{m=-\infty}^\infty e^{im(\varphi-\varphi')}\int_0^\infty k\cos[k(z-z')]\times$$

$$\times \left[I_m'(k\rho) - K_m'(k\rho)\frac{I_m(ka)}{K_m(ka)}\right]K_m(k\rho')dk\Bigg\} \ , \quad (4.47)$$

$$E_\rho(\rho > \rho') = -\frac{q}{2\pi^2\varepsilon_0}\Bigg\{\sum_{m=-\infty}^\infty e^{im(\varphi-\varphi')}\int_0^\infty k\cos[k(z-z')]\times$$

$$\times \left[I_m(k\rho') - K_m(k\rho')\frac{I_m(ka)}{K_m(ka)}\right]K_m'(k\rho)dk\Bigg\} \ , \quad (4.48)$$

$$E_\varphi = \frac{q}{\pi^2\varepsilon_0\rho}\Bigg\{\sum_{m=1}^\infty m\sin[m(\varphi-\varphi')]\int_0^\infty \cos[k(z-z')]\times$$

$$\times \left[I_m(k\rho_<) - K_m(k\rho_<)\frac{I_m(ka)}{K_m(ka)}\right]K_m(k\rho_>)dk\Bigg\} \ , \quad (4.49)$$

$$E_z = \frac{q}{2\pi^2\varepsilon_0}\Bigg\{\sum_{m=-\infty}^\infty e^{im(\varphi-\varphi')}\int_0^\infty k\sin[k(z-z')]\times$$

$$\times \left[I_m(k\rho_<) - K_m(k\rho_<) \frac{I_m(ka)}{K_m(ka)} \right] K_m(k\rho_>) dk \right\} . \tag{4.50}$$

The zeroth order force $\vec{F}_0 = q\vec{E}_0(\vec{r}')$ acting upon the charge q is given by Eq. (4.47) at $\vec{r} = \vec{r}'$ without the first term between brackets (which is the field generated by the charge q itself). There is only a radial component:

$$\vec{F}_0(\vec{r}') = \frac{q^2}{2\pi^2 \varepsilon_0} \left[\sum_{m=-\infty}^{\infty} \int_0^\infty k K_m(k\rho') K_m'(k\rho') \frac{I_m(ka)}{K_m(ka)} dk \right] \hat{\rho}$$

$$= -\frac{q^2}{4\pi^2 \varepsilon_0 \rho'^2} \left\{ \sum_{m=-\infty}^{\infty} \int_0^\infty K_m^2(x) \frac{d}{dx} \left[x \frac{I_m(ax/\rho')}{K_m(ax/\rho')} \right] dx \right\} \hat{\rho} . \tag{4.51}$$

In the last equation we integrated by parts. We plot the zeroth order force of Eq. (4.51) in Fig. 4.3, normalized by the constant $F_q \equiv q^2/4\pi\varepsilon_0 a^2$, as a function of ρ'/a. This force goes to zero when $\rho'/a \to \infty$ and diverges when $\rho' \to a$, as expected.

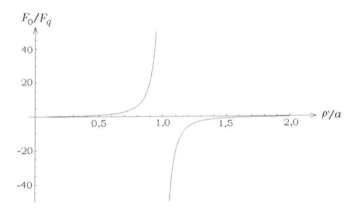

Figure 4.3: Zeroth order force F_0 between an infinite grounded conducting cylinder of radius a and a point charge q at a distance ρ' from the z axis (which is also the axis of symmetry of the cylinder), normalized by the constant $F_q \equiv q^2/4\pi\varepsilon_0 a^2$. For $\rho' < a$ the force comes from Eq. (4.30), while for $\rho' > a$ the force is given by Eq. (4.51).

The surface charges can be calculated using Gauss's law, yielding:

$$\sigma(a, \varphi, z) = \varepsilon_0 E_\rho(a, \varphi, z) =$$

$$= -\frac{q}{2\pi^2 a} \left[\sum_{m=-\infty}^{\infty} e^{im(\varphi-\varphi')} \int_0^\infty \cos[k(z-z')] \frac{K_m(k\rho')}{K_m(ka)} dk \right] . \tag{4.52}$$

The charge per unit length $\lambda(z)$ is given by:

$$\lambda(a, z) = \int_0^{2\pi} \sigma(a, \varphi, z) a \, d\varphi = -\frac{q}{\pi} \int_0^\infty \cos[k(z-z')] \frac{K_0(k\rho')}{K_0(ka)} dk . \tag{4.53}$$

It is interesting to obtain the behaviour of λ for a thin wire, far from z' ($|z-z'| \gg \rho' \gg a$). Utilizing Eq. (3.150) of Jackson's book [13] we obtain:

$$\lambda \approx -\frac{q}{2\ln(|z|/a)} \frac{1}{\sqrt{\rho'^2 + z^2}} . \qquad (4.54)$$

The total charge induced in the cylinder can be obtained integrating Eq. (4.53) from $z = -\infty$ to ∞. Utilizing Eq. (4.33) this yields:

$$Q = \int_{-\infty}^{\infty} \lambda(a,z) dz = -q . \qquad (4.55)$$

A plot of $\lambda(a,z)$ as a function of z, with $z' = 0$ and normalized by q/ρ', is given in Fig. 4.4. The maximum value of $\lambda(a,z)$ is given at $z = z'$, as expected. In Fig. 4.5 we plot $\lambda_{\max}$ as a function of ρ'/a, normalized by q/ρ'. From this Figure we can see that $\lambda_{\max} \to 0$ when $\rho'/a \to \infty$, i.e., for a conducting cylinder of zero thickness, a simple conducting straight line.

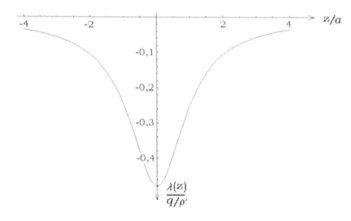

Figure 4.4: Induced linear charge density λ on the conducting cylinder with an external point charge, Eq. (4.53), as a function of z/a. We utilized $z' = 0$, $\rho'/a = 2$ and normalized by q/ρ'.

4.5.2 Thin Cylindrical Shell Held at Zero Potential

Consider that the grounded conducting cylinder is very thin, i.e., $a \ll \rho'$. The modified Bessel functions can be approximated for small argument by [203, Sec. 8.44]:

$$I_m(y \ll 1) \approx \frac{1}{m!} \frac{y^m}{2^m} , \qquad (4.56)$$

$$K_m(y \ll 1) \approx \frac{(m-1)! 2^{m-1}}{y^m}, \quad m > 0 , \qquad (4.57)$$

$$K_0(y \ll 1) \approx -\ln\frac{y}{2} - \gamma . \qquad (4.58)$$

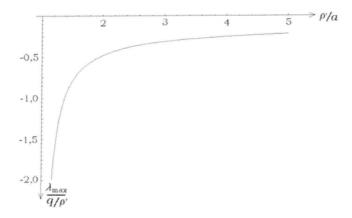

Figure 4.5: Maximum induced linear charge density $\lambda_{\max}(z = z')$ on the conducting cylinder with an external point charge, Eq. (4.53), as a function of ρ'/a. We normalized the plot by q/ρ'.

Here $\gamma = 0.577$ is the Euler-Mascheroni constant.

The term between brackets in Eq. (4.51) for $m = 0$ and for $m > 0$ can be approximated by, respectively:

$$\frac{d}{dx}\left[x\frac{1}{-\ln\frac{ax}{2\rho'} - \gamma}\right] \approx -\frac{1}{\ln(a/\rho')}, \qquad (4.59)$$

$$\frac{d}{dx}\left[x\frac{1}{m!}\frac{(ax/\rho')^m}{2^m}\frac{(ax/\rho')^m}{(m-1)!2^{m-1}}\right] \approx \frac{(2m+1)x^{2m}(a/\rho')^{2m}}{m!(m-1)!2^{2m-1}}. \qquad (4.60)$$

The most relevant term for $\rho' \gg a$ is therefore $m = 0$. Utilizing the identity $\int_0^\infty K_0^2(x)dx = \pi^2/4$ we have the zeroth order force acting upon the charge q as given by:

$$\vec{F}_0(\rho' \gg a) \approx -\frac{q^2}{4\pi^2\varepsilon_0\rho'^2}\int_0^\infty K_0^2(x)\frac{\ln(2\rho'/xa) - \gamma + 1}{[\gamma - \ln(2\rho'/xa)]^2}dx\hat{\rho}$$

$$\approx -\frac{q^2}{4\pi^2\varepsilon_0\rho'^2\ln(\rho'/a)}\int_0^\infty K_0^2(x)dx\hat{\rho} = -\frac{q^2}{16\varepsilon_0\rho'^2\ln(\rho'/a)}\hat{\rho}. \qquad (4.61)$$

Alternatively, another expression for the force can be found by integrating the force exerted by the linear charge density of a thin cylinder, $\lambda(a, z)$ of Eq. (4.53), acting upon the point charge q. Utilizing that

$$\int_{-\infty}^\infty \frac{\rho'\cos[k(z-z')]}{[\rho'^2 + (z-z')^2]^{3/2}}dz = 2kK_1(k\rho'), \qquad (4.62)$$

we obtain:

$$\vec{F}_0(\vec{r}') = -\frac{q}{4\pi\varepsilon_0}\int_{-\infty}^\infty \frac{\rho'}{\sqrt{\rho'^2 + z^2}}\frac{\lambda(a, z)}{\rho'^2 + z^2}dz\hat{\rho}$$

$$= -\frac{q^2}{2\pi^2\varepsilon_0\rho'^2}\int_0^\infty x\frac{K_0(x)K_1(x)}{K_0(xa/\rho')}dx\hat{\rho} \ . \tag{4.63}$$

To compare Eqs. (4.61) and (4.63) we can expand the latter using the approximation $\rho' \gg a$. Utilizing that $K_1(x) = -dK_0/dx$, integrating by parts, and using $K_0(xa/\rho') \approx -\ln(xa/2\rho') - \gamma \approx \ln(\rho'/a)$ we obtain:

$$\vec{F}_0 = \frac{q^2}{2\pi^2\varepsilon_0\rho'^2}\int_0^\infty x\frac{K_0(x)(dK_0/dx)}{K_0(xa/\rho')}dx\hat{\rho}$$

$$\approx -\frac{q^2}{4\pi^2\varepsilon_0\rho'}\int_0^\infty K_0^2(x)\frac{\ln(2\rho'/xa) - \gamma + 1}{[\gamma - \ln(2\rho'/xa)]^2}dx\hat{\rho}$$

$$\approx -\frac{q^2}{16\varepsilon_0\rho'^2\ln(\rho'/a)}\hat{\rho} \ , \tag{4.64}$$

which is exactly Eq. (4.61).

4.5.3 Infinite Cylindrical Shell Held at Constant Potential

Suppose that the conducting cylinder is held at a constant potential, $\phi(a,\varphi,z) = \phi_0$. From Eq. (4.41) we obtain (with $n'' = \rho_<$ and $\rho_> = \rho$):

$$\left.\frac{\partial G}{\partial n''}\right|_{\rho''=a} = -\frac{2}{\pi}\Bigg\{\sum_{m=-\infty}^\infty e^{im(\varphi-\varphi'')}\int_0^\infty k\cos[k(z-z'')]\frac{K_m(k\rho)}{K_m(ka)}\times$$

$$\times \left[I_m'(k\rho'')K_m(ka) - I_m(ka)K_m'(k\rho'')\right]dk\Bigg\}_{\rho''=a}$$

$$= \frac{2}{\pi a}\Bigg\{\sum_{m=-\infty}^\infty e^{im(\varphi-\varphi'')}\int_0^\infty \cos[k(z-z'')]\frac{K_m(k\rho)}{K_m(ka)}dk\Bigg\} \ . \tag{4.65}$$

In the last equality we utilized the Wronskian relation $W[I_m(k\rho''), K_m(k\rho'')] = -1/(k\rho'')$.

The second term given by Eq. (4.8) can be written as:

$$\phi^+ = -\frac{1}{4\pi}\oint_S \phi(\vec{r}'')\frac{\partial G}{\partial n''}da'' = \frac{\phi_0}{2\pi^2 a}\int_{-\infty}^\infty a\,dz''\int_0^{2\pi} d\varphi'' \times$$

$$\times \Bigg\{\sum_{m=-\infty}^\infty e^{im(\varphi-\varphi')}\int_0^\infty \cos[k(z-z')]\frac{K_m(k\rho)}{K_m(ka)}dk\Bigg\}$$

$$= \frac{\phi_0}{\pi}\int_{-\infty}^\infty dz''\int_0^\infty \cos[k(z-z')]\frac{K_0(k\rho)}{K_0(ka)}dk$$

$$= \frac{2\phi_0}{\pi}\int_0^\infty dz''\int_0^\infty \cos[k(z-z')]\frac{K_0(k\rho)}{K_0(ka)}dk \ . \tag{4.66}$$

In the last equality we changed the limits of the integral over z''.

In order to calculate the last integral, we utilize Eq. (4.23). Changing variables, we have:

$$\int_0^\infty \cos[k(z-z')]dz' = \pi\delta(k) . \quad (4.67)$$

The approximation for small argument of $K_0(y)$, namely, $K_0(y) \approx -\ln y$, is in this case inappropriate, because the term $\lim_{k\to 0} K_0(k\rho)/K_0(ka) \to 1$ for any ρ. This is true for an infinite cylinder, but gives no physical insight into the behaviour of the potential as a function of ρ. We should use instead $k \ll 1/\rho < 1/a$, yielding:

$$\phi^+ \approx \phi_0 \frac{\ln(k\rho)}{\ln(ka)}, \quad \text{for} \quad k \ll 1/\rho < 1/a . \quad (4.68)$$

The potential outside an infinite conducting cylinder with an external charge, held at a constant potential ϕ_0, is then given by the summation of Eqs. (4.43) and (4.68).

We can find the potential of a cylinder held at a constant potential ϕ_0 by a different method. Suppose we have a long straight line of length ℓ along the z axis, uniformly charged with a linear charge density λ. The potential at a distance ρ from the z axis, for $\ell \gg \rho$, is given by:

$$\phi_{\text{line}} \approx \frac{\lambda}{2\pi\varepsilon_0} \ln \frac{\ell}{\rho} . \quad (4.69)$$

At a distance $\rho = a$ from the z axis, we have a constant potential $\phi_0 = 2\lambda \ln(\ell/a)$, which is the same boundary condition as before. This implies that the solution is the same. Substituting λ, we obtain the potential as given by:

$$\phi_{\text{line}} = \phi_0 \frac{\ln(\ell/\rho)}{\ln(\ell/a)} . \quad (4.70)$$

Note that Eq. (4.68) with $k \ll 1/\rho < 1/a$ and Eq. (4.70) with $\ell \gg a > \rho$ are essentially the same. Henceforth, we utilize Eq. (4.70) as the solution for a long cylinder held at a constant potential.

The final potential of the problem of a long conducting cylinder held at a constant potential ϕ_0 with an external charge q is given by:

$$\phi(\vec{r},\vec{r}') = \frac{q}{2\pi^2\varepsilon_0} \left\{ \sum_{m=-\infty}^{\infty} e^{im(\varphi-\varphi')} \int_0^\infty \cos[k(z-z')] K_m(k\rho_>) \times \right.$$

$$\left. \times \left[I_m(k\rho_<) - \frac{I_m(ka)}{K_m(ka)} K_m(k\rho_<) \right] dk \right\} + \phi_0 \frac{\ln(\ell/\rho)}{\ln(\ell/a)} . \quad (4.71)$$

The components of the zeroth order electric field $\vec{E}_0$, the zeroth order force $\vec{F}_0$ exerted on q, the surface charge density σ and the linear charge density λ are given by, respectively:

$$E_\rho(\rho < \rho') = -\frac{q}{2\pi^2\varepsilon_0} \left\{ \sum_{m=-\infty}^{\infty} e^{im(\varphi-\varphi')} \int_0^\infty k\cos[k(z-z')] \times \right.$$

$$\times \left[I_m{}'(k\rho) - \frac{I_m(ka)}{K_m(ka)} K'_m(k\rho) \right] K_m(k\rho')dk \bigg\} , \qquad (4.72)$$

$$E_\rho(\rho > \rho') = -\frac{q}{2\pi^2\varepsilon_0} \bigg\{ \sum_{m=-\infty}^{\infty} e^{im(\varphi-\varphi')} \int_0^\infty k\cos[k(z-z')] \times$$

$$\times \left[I_m(k\rho') - \frac{I_m(ka)}{K_m(ka)} K_m(k\rho') \right] K_m{}'(k\rho)dk \bigg\} + \frac{\phi_0}{\rho\ln(\ell/a)} , \qquad (4.73)$$

$$E_\varphi = \frac{q}{\pi^2\varepsilon_0\rho} \bigg\{ \sum_{m=1}^{\infty} m\sin[m(\varphi-\varphi')] \int_0^\infty \cos[k(z-z')] \times$$

$$\times \left[I_m(k\rho_<) - \frac{I_m(ka)}{K_m(ka)} K_m(k\rho_<) \right] K_m(k\rho_>)dk \bigg\} , \qquad (4.74)$$

$$E_z = \frac{q}{2\pi^2\varepsilon_0} \bigg\{ \sum_{m=-\infty}^{\infty} e^{im(\varphi-\varphi')} \int_0^\infty k\sin[k(z-z')] \times$$

$$\times \left[I_m(k\rho_<) - \frac{I_m(ka)}{K_m(ka)} K_m(k\rho_<) \right] K_m(k\rho_>)dk \bigg\} , \qquad (4.75)$$

$$\vec{F}_0(\vec{r}') = -\frac{q^2}{4\pi^2\varepsilon_0\rho'^2} \bigg\{ \sum_{m=-\infty}^{\infty} \int_0^\infty K_m^2(x) \frac{d}{dx}\left[x \frac{I_m(ax/\rho')}{K_m(ax/\rho')} \right] dx \bigg\} \hat{\rho}$$

$$+ \frac{q\phi_0}{\rho'\ln(\ell/a)}\hat{\rho} , \qquad (4.76)$$

$$\sigma(a,\varphi,z) = -\frac{q}{2\pi^2} \bigg\{ \sum_{m=-\infty}^{\infty} e^{im(\varphi-\varphi')} \int_0^\infty k\cos[k(z-z')] \frac{K_m(k\rho')}{K_m(ka)} dk \bigg\}$$

$$+ \frac{\varepsilon_0\phi_0}{a\ln(\ell/a)} , \qquad (4.77)$$

$$\lambda(a,z) = -\frac{q}{\pi} \int_0^\infty \cos[k(z-z')] \frac{K_0(k\rho')}{K_0(ka)} dk + \frac{2\pi\varepsilon_0\phi_0}{\ln(\ell/a)} . \qquad (4.78)$$

From Eq. (4.78) we can calculate the total charge on the cylinder:

$$Q = \int_{-\infty}^{\infty} \lambda(a,z)dz = -\frac{q}{\pi} \int_{-\infty}^{\infty} dz \int_0^\infty \cos[k(z-z')] \frac{K_0(k\rho')}{K_0(ka)} dk + \frac{2\pi\ell\varepsilon_0\phi_0}{\ln(\ell/a)}$$

$$= -q \lim_{k\to 0} \frac{\ln(k\rho')}{\ln(ka)} + \frac{2\pi\ell\varepsilon_0\phi_0}{\ln(\ell/a)} = -q + \frac{2\pi\ell\varepsilon_0\phi_0}{\ln(\ell/a)} . \qquad (4.79)$$

For a neutral charged cylinder, i.e., $Q = 0$, we can relate the constant potential ϕ_0 with the charge q by:

$$\phi_0 = \frac{q\ln(\ell/a)}{2\pi\varepsilon_0\ell} . \qquad (4.80)$$

4.6 Discussion

We can express the zeroth order force exerted by the grounded conducting infinite cylinder of radius a upon the external point charge q at a distance ρ' from the axis of the cylinder as given by:

$$\vec{F}_0 = -\alpha_L \frac{q^2}{4\pi\varepsilon_0 \rho'^2} \hat{\rho} \;, \tag{4.81}$$

where α_L is a dimensionless parameter. In this work we have obtained three different expressions for this force, namely, Eqs. (4.51), (4.61) and (4.63). The parameter α_L for these three cases is given by, respectively:

$$\alpha_L = \frac{1}{\pi} \left\{ \sum_{m=-\infty}^{\infty} \int_0^\infty K_m^2(x) \frac{d}{dx} \left[x \frac{I_m(ax/\rho')}{K_m(ax/\rho')} \right] dx \right\}, \tag{4.82}$$

$$\alpha_L \approx \frac{1}{\pi} \int_0^\infty K_0^2(x) \frac{\ln(2\rho'/xa) - \gamma + 1}{[\gamma - \ln(2\rho'/xa)]^2} dx \approx \frac{\pi}{4\ln(\rho'/a)} \;, \tag{4.83}$$

$$\alpha_L \approx \frac{2}{\pi} \int_0^\infty x \frac{K_0(x) K_1(x)}{K_0(xa/\rho')} dx \;. \tag{4.84}$$

We plot these three values of α_L as functions of a/ρ' in Figs. 4.6 to 4.8.

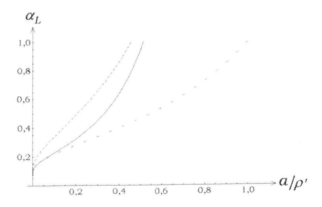

Figure 4.6: Dimensionless parameter α_L given by Eq. (4.81) as a function of a/ρ'. The continuous line represents the parameter from Eq. (4.82); the tight-dashed line that of Eq. (4.83); and the light-dashed line that of Eq. (4.84).

We can see that these three values of α_L converge to one another as $a/\rho' \to 0$. This was expected because Eq. (4.51) is valid for a cylinder of finite thickness with arbitrary value of a/ρ', while Eqs. (4.61) and (4.63) are valid only for a thin cylinder, i.e., for $a \ll \rho'$.

In table (4.85) we present the values of the exact α_L given by Eq. (4.82) as a function of ρ'/a.

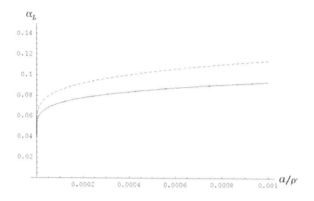

Figure 4.7: Dimensionless parameter α_L given by Eq. (4.81) as a function of a/ρ', for the region $a/\rho' \ll 1$. The continuous line represents the parameter from Eq. (4.82); the tight-dashed line represents the parameter of Eq. (4.83); and the light-dashed line (which in this interval of a/ρ' is overlaid on the continuous line) represents the parameter of Eq. (4.84).

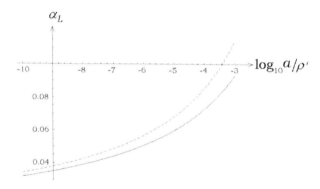

Figure 4.8: Dimensionless parameter α_L given by Eq. (4.81) as a function of $\log_{10}(a/\rho')$, for the region $a/\rho' \ll 1$. The continuous line represents the parameter from Eq. (4.82); the tight-dashed line represents the parameter of Eq. (4.83); and the light-dashed line (which in this interval of a/ρ' is overlaid on the continuous line) represents the parameter of Eq. (4.84).

From Eq. (4.83) we can see that when $a/\rho' \ll 1$, the parameter α_L behaves as $\pi/[4\ln(\rho'/a)]$. That is, it goes to zero when $a/\rho' \to 0$. According to these calculations we conclude that there is no force between a point charge and an idealized grounded conducting line (of zero thickness). One of the authors (AKTA) [1] had expected $0 < \alpha_L < 1$, not specifically for a grounded conducting line, but for a conducting line with zero total charge. In particular he expected that $0.1 < \alpha_L < 0.9$, by guessing the result based on dimensional analysis and in analogy with the case of a point charge q at a distance ρ' from an infinite conducting plane. In this last case, the net force upon the test charge is given by $\alpha_P q^2/4\pi\varepsilon_0 \rho'^2$, with $\alpha_P = 1/4 = 0.25$. The results of the calculations presented here, on the other hand, indicate that $\alpha_L = 0$ when $a/\rho' = 0$ (in the case of a grounded infinite line). This is an interesting result indicating that the existence of a force upon the external test charge requires not only that it is at a finite distance to the cylinder, but also the existence of a surface area different from zero in the conductor with which it is interacting.

$$\begin{bmatrix} \rho'/a & \alpha_L \\ 1.1 & 29.1 \\ 1.2 & 8.94 \\ 1.5 & 2.20 \\ 2.0 & 0.944 \\ 5 & 0.322 \\ 10 & 0.228 \\ 100 & 0.130 \\ 10^3 & 0.0930 \\ 10^4 & 0.0727 \\ 10^{10} & 0.0318 \end{bmatrix} \quad (4.85)$$

Later on we compare this zeroth order force with the force proportional to the voltage of the battery arising when a constant current flows upon the cylindrical wire.

Chapter 5

Relevant Topics

5.1 Properties of the Electrostatic Field

We present here some properties of the scalar electric potential, ϕ, and of the electric field, $\vec{E}$. These properties are proved in detail in most books dealing with electromagnetism, so that we present here only the main aspects. Suppose that we are in an inertial frame of reference S with origin 0. There are N point charges q_j at rest in this reference frame, with $j = 1, ..., N$. The position vector describing the location of charge q_j relative to 0 is represented by $\vec{r}_j$. The electric potential at the point $\vec{r}_o$ due to these N charges, according to the principle of superposition, is defined by:

$$\phi(\vec{r}_o) \equiv \sum_{j=1}^{N} \frac{q_j}{4\pi\varepsilon_0} \frac{1}{r_{oj}} , \qquad (5.1)$$

where $r_{oj} \equiv |\vec{r}_o - \vec{r}_j|$ is the distance between the tip of the vector $\vec{r}_o$ and the charge q_j.

The electric field at the point $\vec{r}_o$ is given by

$$\vec{E}(\vec{r}_o) = -\nabla_o \phi . \qquad (5.2)$$

Performing the line integral between points A and B of the potential difference, $d\phi = -\vec{E} \cdot d\vec{\ell}$, yields

$$\int_A^B \vec{E} \cdot d\vec{\ell} = -\int_A^B (\nabla \phi) \cdot d\vec{\ell} = -\int_A^B d\phi = \phi(\vec{r}_A) - \phi(\vec{r}_B) . \qquad (5.3)$$

That is, this integral is independent of the path of integration, being a function only of the initial and final points.

If it is performed an integration around a closed path of arbitrary form, this yields a null value:

$$\oint \vec{E} \cdot d\vec{\ell} = 0 . \qquad (5.4)$$

5.2 The Electric Field in Different Points of the Cross-section of the Wire

Let us suppose a rectilinear, resistive and homogeneous wire, of uniform cross-section, conducting a steady current. It seems that Davy was the first to prove in 1821 that the current flows over the whole cross-section and not only along the surface of the wire [63, p. 90]:

> As we have already seen, Cavendish investigated very completely the power of metals to conduct electrostatic discharges; their power of conducting voltaic currents was now examined by Davy.[1] His method was to connect the terminals of a voltaic battery by a path containing water (which it decomposed), and also by an alternative path consisting of the metallic wire under examination. When the length of the wire was less than a certain quantity, the water ceased to be decomposed; Davy measured the lengths and weights of wires of different materials and cross-sections under these limiting circumstances; and, by comparing them, showed that the conducting power of a wire formed of any one metal is inversely proportional to its length and directly proportional to its sectional area, but independent of the shape of the cross-section.[2] The latter fact, as he remarked, showed that voltaic currents pass through the substance of the conductor and not along its surface.

A theoretical proof that the current fills the cross-section of the wire can be found in the book of Chabay and Sherwood [166, Section 18.2.4, p. 631]. Suppose there is a solid, homogeneous, rectilinear and uniformly resistive wire, with a cross-section of arbitrary form, carrying a steady current. In steady-state the electric field must be parallel to the wire (to avoid transverse currents and transverse electrostatic polarizations). Imagine now a rectangular path $ABCDA$ within the wire, with AB and CD parallel to the wire, while BC and DA are perpendicular to the wire. When we perform the line integration of the electric field, we get a null value, as was shown in Section 5.1. In this proof we utilized charges at rest. In the case of this Section we are considering steady currents, so that the surface charges are also moving with a drifting velocity of value v_d relative to the bulk of the wire. But these drifting velocities are very small compared to light velocity c. This means that the corrections of second order, of the type v_d^2/c^2, will be negligible in comparison to Coulomb's force. Therefore, they will not be considered here. This means that the electric field in the section AB must be parallel to the wire, with its intensity equal to the intensity of the electric field in the section CD. By the microscopic form of Ohm's law we find that the same result must be true for the volume current density, $\vec{J}$.

---

[1] *Phil. Trans.* cxi (1821), p. 425. His results were confirmed by Becquerel, *Annales de Chimie*, xxxii (1825), p. 423.
[2] These results had been known to Cavendish.

Even with the component of the electric field arising from the radial Hall effect, to be discussed in Section 6.4, the same result will be maintained. The reason for this is that the component of the electric field pointing toward the axis of the wire will have its line integral cancelled between the sections BC and DA.

Utilizing the same reasoning in the case of a circuit having the shape of a solid and homogeneous ring, conducting an azimuthal current, we find that the azimuthal electric field (neglecting the small radial Hall effect) must decrease as $1/\rho$, where ρ is the distance of the observation point to the ring axis of symmetry. We will see an example of this fact in Chapter 13.

We do not know any experience which tried to verify if the electric field and volume current density are really constants in all points of the cross-section of a metallic rectilinear wire carrying a steady current. The same can be said of the $1/\rho$ dependence in the case of a ring. But the experiences of Bergman, Schaefer, Jefimenko and Parker (see Chapter 3) show qualitatively that these suppositions are reasonable.

5.3 Electromotive Force Versus Potential Difference

In this book we will see several examples showing that the electric field outside a wire carrying a steady current is proportional to the electromotive force (emf) of the battery connected to the wire. In this Section we emphasize that the emf is a concept different from the potential difference due to charges at rest. This topic has been discussed by a number of authors [204] [16, Section 7.1.2, pp. 277-278] [171] [166, pp. 642-644].

In order to separate positive and negative charges it is necessary the existence of "non-Coulomb" forces, $\vec{F}_{nC}$, i.e., forces which are not of electrostatic origin. This must happen in all cases in which we separate these charges: in frictional electricity; when two different metals touch one another; in a chemical battery; in thermoelectric effect; in piezoelectric effect; in a Van de Graaff generator; in a photoelectric cell, *etc*. The reason for this is that due to Coulomb's force, charges of opposite sign attract one another and tend to get together. What separate these charges (or prevent them from getting together once they were separated) can then only be of non-electrostatic origin. That is, this interaction must be independent from Coulomb's electrostatic force.

The origin of the expression "electromotive force" is due to Volta (1745-1827) [204].

As we have seen in Section 5.1, the difference of electrostatic potential between two points A and B is given by

$$\phi_B - \phi_A = -\int_A^B \vec{E}_C \cdot d\vec{\ell}. \qquad (5.5)$$

Here $\vec{E}_C$ is the electrostatic field due to charges at rest. This potential difference does not depend upon the path of integration, being a function only of the initial and final points.

On the other hand, the electromotive force between two points A and B, emf_{BA}, is given by

$$emf_{BA} = \int_A^B \vec{E}_{nC} \cdot d\vec{\ell} \,. \tag{5.6}$$

Here $\vec{E}_{nC} = \vec{F}_{nC}/q$ is the impressed force acting upon the test charge q, divided by the value of this charge. This impressed force has a non-electrostatic origin. This line integral depends upon the path of integration.

In a battery, for instance, the Coulomb and non-Coulomb forces balance one another in an open circuit. This means that there will be a potential difference across the battery. Moreover, this potential difference is numerically equal to the battery's emf. The emf of a battery is also called its voltage.

Analogously, the emf of a closed circuit is given by:

$$emf = \oint \vec{E}_{nC} \cdot d\vec{\ell} \,. \tag{5.7}$$

If in the closed circuit there is a chemical battery, or another force of non-electrostatic origin, this line integral upon a closed circuit may have a net value different from zero.

Despite the term "force" in the expression emf, the emf is not a force in the Newtonian sense. The emf of a pile or chemical battery is numerically equal to the potential difference generated between the terminals of the battery. It has the same units as potential difference, namely, volt or newton/coulomb. Despite this fact, the emf is not a potential difference, as we emphasized in this Section. Its origin is due to a non-electrostatic force. And it is not always associated with a potential difference. For instance, in the case of a ring approaching or moving away from a permanent magnet (example discussed by Weber, as we discuss in Appendix A), it is generated a current along the resistive ring, although there is no potential difference between any two points of the ring [204] [184].

5.4 Russell's Theorem

Russell proved in an important short paper a general theorem related with straight parallel conductors of arbitrary cross-sections carrying steady currents [9]. He concluded that the density of surface charges σ on the conductors vary linearly with distance along the direction of their common axis z. The same was found valid for the potential ϕ inside and outside the conductors.

He considered homogeneous isotropic materials surrounded by an insulating medium of constant permittivity ε. His theorem is valid at great distances from their termination (in order to neglect edge effects) and also far from the sources of electromotive force, emf, maintaining the current.

The essence of his proof is to consider that inside the conductors carrying steady currents the electric field $\vec{E}$ has everywhere the same longitudinal component. By $\vec{E} = -\nabla\phi$ this means that the potential inside them must be a linear function of z. Outside the conductors the potential ϕ must satisfy Laplace's equation $\nabla^2\phi = 0$. But the solutions of Laplace's equation which satisfy all the boundary conditions are unique. As the boundary conditions in all conductors are linear functions of z, the same must be true outside them. That is,

$$\phi(x,y,z) = F(x,y)(A + Bz) , \qquad (5.8)$$

where $F(x,y)$ is a function of the transverse coordinates, while A and B are constants.

Analogously, the surface charge densities σ are obtained by Gauss's law as directly proportional to the normal component of the electric field at the conductor boundaries. As $\vec{E} = -\nabla\phi$, Eq. (5.8) yields:

$$\sigma(x,y,z) = G(x,y)(A + Bz) , \qquad (5.9)$$

where $G(x,y)$ is a function of the transverse coordinates.

This means that the solution of electrostatic problems can be directly applied to the solution of steady currents, by including a linear dependence in the longitudinal component. In the following Chapters we will see many examples illustrating this theorem.

But it should be kept in mind that it is valid only far from the terminations of the conductors and also far from the batteries. Moreover, it is not valid as well close to the junction of two materials of different conductivities.

Part II
Straight Conductors

In this work the frame of reference will always be the laboratory. When we speak of conductors and wires in general, it should be understood that they are usually uniformly resistive, unless stated otherwise. The medium outside the conductors or between them will be usually air or vacuum. No time variation of currents or potentials will be considered here. It is assumed that there are no conductors nor other external charges close to the current-carrying wire, so that we will consider it isolated from external influences (except for the test charge already mentioned).

In this first part we will consider one or more parallel resistive conductors carrying steady currents along the z axis.

Chapter 6

A Long Straight Wire of Circular Cross-section

To our knowledge the first to perform theoretical calculations related to the electric field inside a wire of circular cross section due to surface charges increasing linearly with the longitudinal coordinate has been Wilhelm Weber in 1852 [32], as we discuss in the Appendix A. Here we follow the treatment published in 1999 [1].

6.1 Configuration of the Problem

The situation considered here is that of a cylindrical and homogeneous resistive wire of length ℓ and radius $a \ll \ell$, Figure 6.1.

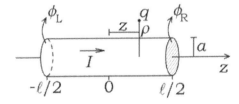

Figure 6.1: Configuration of the problem.

The axis of the wire coincides with the z direction, with $z = 0$ at the center of the wire. A battery maintains constant potentials ϕ_L and ϕ_R at the extremities $z = -\ell/2$ and $z = +\ell/2$ of the wire, respectively. The wire carries a constant current I, has a finite conductivity g and is at rest relative to the laboratory. There is air or vacuum outside the wire. At a distance $\rho = \sqrt{x^2 + y^2}$ from the axis of the wire there is a stationary point charge q. We want to know the force exerted by the wire upon the charge q. In particular we wish to calculate the component of this force which is proportional to the voltage or emf of the battery, or to the potential difference acting along the wire. To this end we will suppose the following approximation:

$$\ell \gg \rho \geq 0, \ \ell \gg a > 0 \text{ and } \ell \gg |z| \geq 0 \ . \tag{6.1}$$

Here z is the longitudinal component of the vector position of q. See Fig. 6.1. We utilize throughout this chapter cylindrical coordinates (ρ, φ, z) and unit vectors $\hat{\rho}$, $\hat{\varphi}$ and $\hat{z}$.

This wire must be closed somewhere. The calculations presented here with this approximation should be valid for the circuit of Figure 6.2. This is a square circuit with four sides of length ℓ, composed of cylindrical wires of radius $a \ll \ell$. There is an external point charge close to the middle of one of its sides (like AB, BC or CD) and far from the battery. The case when the point charge is close to the middle of the side AD has been published in 2004 [205].

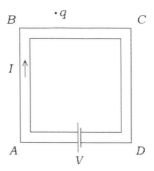

Figure 6.2: A closed square circuit made of a resistive wire of circular cross-section. There is a point charge close to the middle of one of its sides.

With this approximation we can consider that the three other sides will not contribute significantly to the potential and field near the center of the fourth side. Alternatively, it should also give approximate results for a circular loop of larger radius $R_0 = \ell/2\pi$ and smaller radius $a \ll R_0$ (a ring), if the point charge is close to the ring but far from the battery maintaining the current. It might even be utilized as a first gross approximation for the force on the point charge of Figure 1.1 considering a generic circuit of large length and small curvatures (that is, with radii of curvature much larger than the diameter of the wire and also much larger than the distance of the point charge to the wire).

We consider separately three components of the force exerted by the wire on q: (A) The component due to the electrostatic induction, based upon the charges induced along the surface of the wire by q, which has been considered in Chapter 4; (B) the component of the force due to the surface charges which exist in resistive wires carrying steady currents (force proportional to the voltage or emf of the battery connected to the circuit); and (C) the force proportional to the square of the drifting velocity, v_d, of the conduction electrons, i.e., proportional to v_d^2/c^2.

6.2 Force Proportional to the Potential Difference Acting upon the Wire

When a constant current flows in a resistive wire connected to a battery, the electric field driving the conduction electrons against the resistive friction exerted upon them by the lattice is due to free charges distributed along the surface of the wire, as we have seen before. We represent this surface charge density by $\sigma(a, \varphi, z)$. For steady currents, σ is constant in time but varies along the length of the wire (that is, it is a function of z). Here we follow the approach of Weber and Kirchhoff discussed in Chapter 2. The battery, due to the chemical forces which maintain its terminals at different potentials, is responsible for maintaining this distribution of charges along the surface of the wire. But the battery does not generate directly the electric field in all points along the circuit. The surface charges, on the other hand, generate not only the electric field inside the wire but also an electric field outside it.

The approach of this chapter is the following: We consider the cylindrical wire carrying the constant current I and calculate the potential ϕ_1 and electric field $\vec{E}_1$ inside and outside the wire due to these surface charges in the absence of the test charge q. When we put the test charge at a distance ρ from the wire the force on it due to the surface charges will be then given by $\vec{F}_1 = q\vec{E}_1$, supposing that it is small enough such that it does not disturb the current nor the wire (except from the induction charges already considered in Chapter 4, which will exert the force $\vec{F}_0 = q\vec{E}_0$). We begin calculating the potential due to the surface charges.

As there is a constant current in the wire, the electric field inside it and driving the current must be constant over the cross-section of the wire [63, p. 90]. Here we are disregarding the small radial Hall effect inside the wire due to the azimuthal magnetic field generated by the current to be discussed in Section 6.4. This means that the potential and surface charge distribution must be a linear function of z, as we saw in Section 5.4. Due to the axial symmetry of the wire it cannot depend on the azimuthal angle either. This means that

$$\sigma(a, \varphi, z) = \sigma_A + \sigma_B \frac{z}{\ell}, \qquad (6.2)$$

where σ_A and σ_B are constants.

Before proceeding we wish to discuss this expression. We may wish to consider the wire as globally neutral, i.e., no net charge as a whole. When we integrate the free charge density σ over the whole surface of the wire we need to obtain a zero net value in this case. This will happen with Eq. (6.2) after integrating from $z = -\ell/2$ to $z = \ell/2$ only in the symmetrical case in which $\sigma_A = 0$. This might represent, for instance, the top side BC of Figure 6.2. On the other hand, we will perform the calculations with a generic value of σ_A so that the calculation might be applicable, for instance, to the left half of the top side BC of Figure 6.2. The integration of σ over this left side (from $z = -\ell/2$ to zero) will yield a positive value, as it is closer to the positive terminal. This positive charge will be balanced by the negative charge lying on the right half of the top side of Figure 6.2 (z going from zero to $+\ell/2$). With a generic σ_A we might also consider, for instance, the left side AB of Figure 6.2 with a positive charge, which will be balanced by the negative charge in the right side CD of Figure 6.2. It should be emphasized that the point where $\sigma = 0$ is specified by the battery. The battery itself also specifies where σ will be positive (portions of the wire closer to its positive terminal) or negative (portions of the wire closer to its negative terminal).

Due to the axial symmetry of σ we can calculate ϕ at the specific angle $\varphi = 0$ rad and then generalize the solution to all φ. The potential inside or outside the wire is then given by:

$$\phi_1(\rho, z) = \frac{1}{4\pi\varepsilon_0} \int_{\varphi_2=0}^{2\pi} \int_{z_2=-\ell/2}^{\ell/2} \frac{\sigma a \, d\varphi_2 dz_2}{\sqrt{\rho^2 + a^2 - 2\rho a \cos\varphi_2 + (z_2 - z)^2}}$$

$$= \frac{1}{4\pi\varepsilon_0} \int_{\varphi_2=0}^{2\pi} \int_{z_2=-\ell/2}^{\ell/2} \frac{(\sigma_A + \sigma_B z_2/\ell)d\varphi_2 dz_2}{\sqrt{\left(1 - 2\frac{\rho}{a}\cos\varphi_2 + \frac{\rho^2}{a^2}\right) + \left(\frac{z_2-z}{a}\right)^2}} \, . \quad (6.3)$$

Defining the dimensionless variables $s^2 \equiv 1 - 2(\rho/a)\cos\varphi_2 + (\rho^2/a^2)$ and $u \equiv (z_2 - z)/a$ we are then led to: $\phi_1(\rho, z) = (a/4\pi\varepsilon_0)[(\sigma_B a/\ell)I_1 + (\sigma_A + \sigma_B z/\ell)I_2]$, where

$$I_1 \equiv \int_{\varphi_2=0}^{2\pi} \int_{u=-(\ell/2a+z/a)}^{\ell/2a-z/a} u \frac{d\varphi_2 du}{\sqrt{s^2 + u^2}} \, , \quad (6.4)$$

and

$$I_2 \equiv \int_{\varphi_2=0}^{2\pi} \int_{u=-(\ell/2a+z/a)}^{\ell/2a-z/a} \frac{d\varphi_2 du}{\sqrt{s^2 + u^2}} \, . \quad (6.5)$$

These integrals can be solved with the approximation (6.1). The final approximate result is given by (generalizing for all φ):

$$\phi_1(\rho \leq a, \varphi, z) \approx \frac{a}{\varepsilon_0}\left[\sigma_A \ln\frac{\ell}{a} + \sigma_B \frac{z}{\ell}\ln\frac{\ell}{ea}\right] \, , \quad (6.6)$$

$$\phi_1(\rho \geq a, \varphi, z) \approx \frac{a}{\varepsilon_0}\left[\sigma_A \ln\frac{\ell}{\rho} + \sigma_B \frac{z}{\ell}\ln\frac{\ell}{e\rho}\right] \, . \quad (6.7)$$

From Eq. (6.1) we can neglect $\ln e = 1$ in comparison with $\ln(\ell/a)$ and $\ln(\ell/\rho)$. This yields:

$$\phi_1(\rho, \varphi, z) \approx \frac{a\sigma(z)}{\varepsilon_0}\ln\frac{\ell}{a} = \frac{a(\sigma_A + \sigma_B z/\ell)}{\varepsilon_0}\ln\frac{\ell}{a} \, , \text{ if } \rho \leq a \, , \quad (6.8)$$

$$\phi_1(\rho,\varphi,z) \approx \frac{a\sigma(z)}{\varepsilon_0} \ln\frac{\ell}{\rho} = \frac{a(\sigma_A + \sigma_B z/\ell)}{\varepsilon_0} \ln\frac{\ell}{\rho}, \quad \text{if } \rho \geq a. \tag{6.9}$$

By writing the linear density of charges along the wire as $\lambda(z) \equiv 2\pi a\sigma(z)$ these last expressions can be written as

$$\phi_1(\rho \leq a,\varphi,z) \approx \frac{\lambda(z)}{2\pi\varepsilon_0} \ln\frac{\ell}{a}, \tag{6.10}$$

$$\phi_1(\rho \geq a,\varphi,z) \approx \frac{\lambda(z)}{2\pi\varepsilon_0} \ln\frac{\ell}{\rho}. \tag{6.11}$$

From Eqs. (6.6) and (6.7) we can obtain the electric field $\vec{E}_1 = -\nabla\phi_1$:

$$\vec{E}_1(\rho < a,\varphi,z) \approx -\frac{a\sigma_B}{\varepsilon_0\ell}\left(\ln\frac{\ell}{ea}\right)\hat{z}, \tag{6.12}$$

$$\vec{E}_1(\rho > a,\varphi,z) \approx \frac{a}{\varepsilon_0}\left(\sigma_A + \sigma_B\frac{z}{\ell}\right)\frac{\hat{\rho}}{\rho} - \frac{a\sigma_B}{\varepsilon_0\ell}\left(\ln\frac{\ell}{e\rho}\right)\hat{z}. \tag{6.13}$$

To our knowledge the first to obtain Eq. (6.12) beginning with the integration of Eq. (6.2) was Wilhelm Weber in 1852, as we discuss in Appendix A.

From Eq. (6.1) we can neglect 1 in comparison with $\ln(\ell/a)$ and $\ln(\ell/\rho)$. This yields the coulombian force on a test charge q located at (ρ,φ,z) as given by (with $\vec{F}_1 = -q\nabla\phi_1$):

$$\vec{F}_1 = q\vec{E}_1 \approx -\frac{qa}{\varepsilon_0}\frac{\partial\sigma(z)}{\partial z}\left(\ln\frac{\ell}{a}\right)\hat{z} = -\frac{qa\sigma_B}{\ell\varepsilon_0}\left(\ln\frac{\ell}{a}\right)\hat{z} \quad \text{if } \rho < a, \tag{6.14}$$

$$\vec{F}_1 = q\vec{E}_1 \approx \frac{qa\sigma(z)}{\varepsilon_0}\frac{\hat{\rho}}{\rho} - \frac{qa}{\varepsilon_0}\frac{\partial\sigma(z)}{\partial z}\left(\ln\frac{\ell}{\rho}\right)\hat{z}$$

$$= \frac{qa(\sigma_A + \sigma_B z/\ell)}{\varepsilon_0}\frac{\hat{\rho}}{\rho} - \frac{qa\sigma_B}{\ell\varepsilon_0}\left(\ln\frac{\ell}{\rho}\right)\hat{z} \quad \text{if } \rho > a. \tag{6.15}$$

We can relate these expressions with the current I flowing in the wire. From Figure 6.1 and the fact that ϕ_1 is a linear function of z we obtain

$$\phi_1(\rho \leq a, z) = \frac{\phi_R + \phi_L}{2} + (\phi_R - \phi_L)\frac{z}{\ell}. \tag{6.16}$$

Equating this with Eq. (6.8) and utilizing Ohm's law $\phi_L - \phi_R = RI$, where $R = \ell/g\pi a^2$ is the resistance of the wire, with g being its conductivity, yields $\sigma_B = -R\varepsilon_0 I/a\ln(\ell/a)$ and $\sigma_A = \varepsilon_0(\phi_R + \phi_L)/2a\ln(\ell/a) = \varepsilon_0(RI + 2\phi_R)/2a\ln(\ell/a)$. The density of free charges along the surface of the wire can then be written as:

$$\sigma(a, \varphi, z) = \frac{\varepsilon_0(\phi_R + \phi_L)}{2a \ln(\ell/a)} - \frac{R\varepsilon_0 I}{a \ln(\ell/a)} \frac{z}{\ell} . \tag{6.17}$$

This means that the potential and the force on the test charge q are given by:

$$\phi_1 = \frac{\phi_R + \phi_L}{2} - RI\frac{z}{\ell} \quad \text{if } \rho \leq a , \tag{6.18}$$

$$\phi_1 = \frac{\phi_R + \phi_L}{2} \frac{\ln(\ell/\rho)}{\ln(\ell/a)} - RI\frac{\ln(\ell/\rho)}{\ln(\ell/a)} \frac{z}{\ell} \quad \text{if } \rho \geq a , \tag{6.19}$$

$$\vec{F}_1 = q\vec{E}_1 = q\frac{RI}{\ell}\hat{z} \quad \text{if } \rho < a , \tag{6.20}$$

$$\vec{F}_1 = q\vec{E}_1 = q\left[\frac{1}{\ln(\ell/a)}\left(\frac{RI + 2\phi_R}{2} - RI\frac{z}{\ell}\right)\frac{\hat{\rho}}{\rho} + \frac{RI}{\ell}\frac{\ln(\ell/\rho)}{\ln(\ell/a)}\hat{z}\right] \quad \text{if } \rho > a . \tag{6.21}$$

Now that we have obtained the potential outside the wire we might also invert the argument. That is, we might solve Laplace's equation $\nabla^2 \phi = 0$ in cylindrical coordinates inside and outside the wire (for $a \leq \rho \leq \ell$) by the method of separation of variables, supposing a solution of the form $\phi(\rho, \varphi, z) = R(\rho)\Phi(\varphi)Z(z)$. The arbitrary constants obtained by this method are found imposing the following boundary conditions: finite $\phi(0, \varphi, z)$, $\phi(a, \varphi, z) = (\phi_R + \phi_L)/2 + (\phi_R - \phi_L)z/\ell$ and $\phi(\ell, \varphi, z) = 0$. This last condition is not a trivial one and was obtained only after we found the solution in the order presented in this work. See Eq. (6.9). The usual boundary condition that the potential goes to zero at infinity does not work in the case of a long cylinder carrying a steady current. But the potential going to zero at $\rho = \ell$ is a reasonable result. After all, this means that we are considering $\phi = 0$ at a great distance from the wire. By this reverse method we obtain the potential inside and outside the wire, then the electric field by $\vec{E} = -\nabla\phi$ and lastly the surface charge density by ε_0 times the normal component of the electric field outside the wire in the limit in which $\rho \to a$. In this way we checked the calculations.

If we put $\phi_L = \phi_R = \phi_0$ or $I = 0$ in Eqs. (6.18) to (6.21) we recover the electrostatic solution (long wire charged uniformly with a constant charge density σ_A, with total charge $Q_A = 2\pi a \ell \sigma_A$), namely:

$$\phi_1(\rho \leq a) = \phi_0 = \frac{a\sigma_A}{\varepsilon_0} \ln\frac{\ell}{a} , \tag{6.22}$$

$$\phi_1(\rho \geq a) = \phi_0 \frac{\ln(\ell/\rho)}{\ln(\ell/a)} = \frac{a\sigma_A}{\varepsilon_0} \ln\frac{\ell}{\rho} , \tag{6.23}$$

$$\vec{E}_1(\rho < a) = \vec{0} , \tag{6.24}$$

$$\vec{E}_1(\rho > a) = \frac{\phi_0}{\ln(\ell/a)}\frac{\hat{\rho}}{\rho} = \frac{a\sigma_A}{\varepsilon_0}\frac{\hat{\rho}}{\rho} . \tag{6.25}$$

We can also obtain the capacitance per unit length of this long cylindrical wire as $C/\ell = [Q_A/\phi(a)]/\ell = 2\pi\varepsilon_0/\ln(\ell/a)$.

It is interesting to analyze here the solutions for points extremely close to the wire, $\rho = a + d$ with $d \ll a$. In this approximation Eq. (6.9) yields:

$$\phi_1 \approx \frac{a\sigma(z)}{\varepsilon_0}\left(\ln\frac{\ell}{a} - \frac{d}{a}\right). \tag{6.26}$$

The analysis presented here refines the previous work of Coombes and Laue, who discussed in 1981 the limiting case of an infinitely long wire [17]. They arrived at the same uniform electric field both inside and outside the wire. This is correct for an infinitely long wire. In the present case we arrived at a uniform electric field inside the wire and at an electric field outside the wire with longitudinal and radial components depending on ρ, as we were considering a large but finite length ℓ.

Eqs. (6.16), (6.20) and (6.21) show that the electric field both inside and outside the wire is proportional to the potential difference $\phi_L - \phi_R = RI$ acting along the wire. The same can be said of the force exerted upon a stationary external test charge by the resistive wire carrying a steady current. If we change the diameter of the wire, or its resistivity, we can change the resistance of the wire. But if it is connected to the same battery, in such a way that it is under the action of the same potential difference, the current flowing along the wire will change accordingly. But the density of surface charges and the external electric field will not change. This important aspect has been emphasized by Chabay and Sherwood [165, 166, 171].

Moreover, there will be not only a tangential component of the electric field outside the wire (as might be expected from the continuity of this component at an interface between two media), but also a radial component. In the symmetrical case in which $\phi_L = -\phi_R = RI/2$ the ratio of the radial component of $\vec{F}_1$ to its tangential component is given by $z/[\rho\ln(\ell/\rho)]$. For a wire with 1 meter length with $z = \rho = 10$ cm this ratio is given by 0.4. This means that these two components are of the same order of magnitude.

The longitudinal component of the electric field is continuous at an interface separating two media. From Eqs. (6.20) and (6.21) we can see that at the surface of the wire, $\rho = a$, the longitudinal component of the electric field is given by: $\vec{E}_1 = (RI/\ell)\hat{z}$. This electric field will act upon the surface conduction electrons belonging to the wire, so that they will move with a constant tangential velocity in steady state, with the electric force balanced by the Ohmic resistance. In equilibrium there will be the same number of free electrons entering and leaving a circular strip of length $2\pi a$ and width dz, so that the distribution of surface charges will not change with time, although being a function of z. In any event, the surface charges will not remain stationary when there is a steady current, but will move due to this tangential electric field existing at the surface of the conductor. The drifting velocity of the conduction electrons will be different from zero not only in the bulk of the metal, but also along its surface. The distribution of surface charges is actually a surface current. The same will happen for the other resistive conductors carrying steady currents discussed in this work.

Eqs. (6.19) and (6.21) show that the external potential, electric field and force go to zero when $\ell/a \to \infty$. This means that resistive electric currents exert forces upon external static charges, except in the idealized case of filamentary current (zero cross-section conductor). The crucial aspect for the existence of a force is not only that the current-carrying wires are resistive but that they have finite cross-sections.

Below we consider a force due to the square of the current.

6.3 Force Proportional to the Square of the Current

Up to now we have only considered two components of the force exerted by the resistive wire upon the external test charge: (A) the component arising from electrostatic induction (due to induction charges along the surface of the wire generated by the presence of q); and (B) the component arising from the potential difference acting along the wire (due to the charges along the surface of the wire induced by the presence of the battery, when there is a steady current flowing along the wire). This surface charge density and the accompanying electric field are proportional to the emf of the battery or to the potential difference acting along the wire. We have not yet taken into account the force of the stationary lattice and mobile conduction electrons on the stationary test charge. We consider it here in this Section, analyzing two different theoretical models: Lorentz's force and Weber's force.

We first consider Lorentz's force (or Liénard-Schwarzschild's force). In this case there are also components of the force exerted by a charge q_2 belonging to the current carrying circuit on the test charge q which depend on the square of the velocity of q_2, v_d^2, and on its acceleration. If we have a steady current, the acceleration of q_2 will be its centripetal acceleration due to any curvature in the wire, proportional to v_d^2/r_c, where r_c is the radius of curvature of the wire at each point. This might lead to a force proportional to v_d^2 or to I^2. However, it has been shown that if we have a closed circuit carrying a constant current, there is no net effect of the sum of all these terms on a stationary charge outside the wire [11, page 697, exercise 14.13] [15] [23, Section 6.6]. The same result is valid for Clausius's force law. In conclusion we might say the following: According to Lorentz's force, the stationary lattice creates an electric field which is just balanced by the force due to the free electrons inside the closed wire, even when there is a constant current along the resistive wire. This might be interpreted as considering the wire to be electrically neutral in its interior (the radial Hall effect will be considered later on).

We now consider Weber's electrodynamics [23]. As already stated, we are disregarding the small radial Hall effect inside the wire due to the azimuthal magnetic field generated by the current. This means that the interior of the wire can be considered essentially neutral. Despite this fact Weber's electrodynamics predicts a force exerted by this neutral wire in a stationary charge nearby, even for closed circuits carrying constant currents. The reason for this effect is that the force exerted by the mobile electrons on the stationary test charge is different from the force exerted by the stationary positive ions of the lattice on the test charge. One of us has already performed these calculations in related situations, so that we present here only the final result. The calculations have been published in 1991 [22] [23, Section 6.6, pages 161-168]. When we first performed these calculations we were not completely conscious of the surface charges discussed in this book (proportional to the emf of the battery, or to the potential difference acting along the wire). For this reason the calculations were performed supposing wires electrically neutral in all internal points and also along their surfaces. Despite the limitations of this supposition, we reproduce the final results here in order to show that they are different from the final results obtained with Lorentz's force when we assume the same conditions of neutrality.

Once more we assume (6.1). For the situation of Figure 6.1, with a uniform current density $\vec{J} = (I/\pi a^2)\hat{z}$, the force on the test charge is given by:

$$\vec{F}_2 = -q \frac{I v_d}{4\pi\varepsilon_0 c^2} \frac{\hat{\rho}}{\rho} = -\frac{\mu_0}{4\pi^2} \frac{qI^2}{a^2 en} \frac{\hat{\rho}}{\rho} \quad \text{if } \rho > a \; , \tag{6.27}$$

where v_d is the drifting velocity of the electrons. We also utilized $c^2 = 1/\varepsilon_0\mu_0$ and $v_d = I/\pi a^2 en$, where $e = 1.6 \times 10^{-19}$ C is the elementary charge and n is the number of free electrons per unit volume.

This force is proportional to the square of the current. The electric field $\vec{E}_2 = \vec{F}_2/q$ points toward the current, as if the wire had become negatively charged. Sometimes this second order field is called motional electric field.

Suppose that we now bend the wire carrying a constant current (by letting its shape in the form of a ring, for instance). In this case Weber's electrodynamics predicts another component of the force exerted by this current upon a stationary charge outside the wire. This new component depends upon the acceleration of the source charges (in this case conduction electrons). As we are supposing a steady current which does not change with time, the relevant acceleration here is the centripetal one proportional to v_d^2/r_c, where r_c is the radius of curvature of the wire at that location. This means that also this component of the force will be proportional to v_d^2 or to I^2. The order of magnitude is the same as the previous example. In 1991 [22] and in 1994 [23, Section 6.6, pp. 161-168] it was calculated the net second order force acting upon a stationary charge outside the wire due to a circular closed circuit carrying a steady azimuthal current in the shape of a ring, utilizing Weber's force. We showed that its net value had the order of magnitude of Eq. (6.27). To this end we have taken into account not only the component of the force which depends upon the square of the velocity of the source charges, v_d^2, but also the component of the force due to the centripetal acceleration of the source electrons. This means that Weber's second order force does not go to zero even for closed circuits.

In the case of Lorentz's force, on the other hand, this net second order force is always null in the case of closed currents. This is an important theoretical difference between these two theories.

6.4 Radial Hall Effect

Another simple question which might be asked is the following: Is a stationary resistive wire carrying a constant current electrically neutral in its interior?

Many authors quoted in Section 1.2 answered positively to this question as this was one of their reasons for believing that this wire would not generate any electric field outside itself. However, we already showed that there will be a longitudinal distribution of surface charges which will give rise to the longitudinal electric field inside the wire and also to an electric field outside it. Here we show that there will also be a radial electric field inside the wire due to the fact that its interior is negatively charged.

To our knowledge the first to consider this effect and to present quantitative calculations were Matzed, Russell and Rosser [206, 167]. Smythe also discussed this subject briefly [207, Section 6.04, pp. 250-252].

The usual Hall effect is discussed in most textbooks on classical electromagnetism, so that we will not enter into details here. Normally they consider the effects upon a current carrying conductor when placed in an external magnetic field. These effects include the so-called "Hall voltage" and related topics.

However, what we discuss here is a similar effect but due to the internal magnetic field generated by the current-carrying wire itself, without the presence of any external magnetic field. To distinguish this effect from the usual Hall effect, we utilize the expression radial Hall effect (related to the case of a current flowing along a cylindrical conductor).

We here consider the radial Hall effect due to the azimuthal magnetic field inside the wire generated by the longitudinal current flowing in this wire. As is usually considered [63, p. 90], we will suppose the constant total current I to flow uniformly over the cross-section of the cylindrical wire with a current density $J = I/\pi a^2$. With the magnetic circuital law $\oint_C \vec{B} \cdot d\vec{\ell} = \mu_0 I_C$, where C is the circuit of integration and I_C is the current passing through the surface enclosed by C, we obtain that the magnetic field inside and outside the wire is given by:

$$\vec{B}(\rho \leq a) = \frac{\mu_0 I \rho}{2\pi a^2}\hat{\varphi} , \qquad (6.28)$$

$$\vec{B}(\rho \geq a) = \frac{\mu_0 I}{2\pi \rho}\hat{\varphi} . \qquad (6.29)$$

The magnetic force on a specific conduction electron of charge $q = -e$ inside the wire (due to the magnetic field generated by all other conduction electrons), at a distance $\rho < a$ from the axis and moving with drifting velocity $\vec{v} = -|v_d|\hat{z}$ is given by:

$$\vec{F} = q\vec{v} \times \vec{B} = -\frac{|\mu_0 e v_d I \rho|}{2\pi a^2}\hat{\rho}. \tag{6.30}$$

This radial force pointing inwards will create a concentration of negative charges in the body of the conductor. This is like a pinch effect. In equilibrium there will be a radial force generated by these charges which will balance the magnetic force: $qE = qvB$. That is, there will be inside the wire, beyond the longitudinal electric field E_1 driving the current, a radial electric field pointing inwards given by:

$$\vec{E}_\rho(\rho \leq a) = -\frac{|\mu_0 v_d I \rho|}{2\pi a^2}\hat{\rho}. \tag{6.31}$$

The longitudinal electric field inside the wire driving the current is given by $E_1 = RI/\ell$. In order to compare it with the magnitude of the radial electric field E_ρ due to the Hall effect we consider the maximum value of this last field very close to the surface of the wire, at $\rho \to a$: $E_\rho \to |\mu_0 v_d I|/2\pi a$. This means that (with $R = \ell/g\pi a^2$):

$$\frac{|E_\rho|}{|E_1|} = \frac{|\mu_0 v_d g a|}{2}. \tag{6.32}$$

For a typical copper wire ($v_d \approx 4 \times 10^{-3}$ m/s and $g = 5.7 \times 10^7$ $\Omega^{-1}\text{m}^{-1}$) with 1 mm diameter this yields: $E_\rho/E_1 \approx 7 \times 10^{-5}$. This shows that the radial electric field inside the wire is negligible compared to the longitudinal one.

By Gauss's law $\nabla \cdot \vec{E} = \rho_c/\varepsilon_0$ we obtain that inside the wire there will be a constant negative charge density ρ_{c-} given by: $\rho_{c-} = -|Iv_d|/\pi a^2 c^2$. The total charge inside the wire is compensated by a positive charge spread over the surface of the wire with a constant surface density $\sigma_+ = |\rho_{c-}a/2| = |Iv_d|/2\pi a c^2$. That is, the negative charge inside the wire in a small segment of length dz, $\rho_{c-}\pi a^2 dz$, is equal and opposite to the positive charge along its surface, $\sigma_+ 2\pi a dz$. This means that the radial Hall effect will not generate any electric field outside the wire, only inside it. For this reason it is not relevant to the experiments discussed before. In any event it is important to clarify this effect.

Contrary to the surface density of free charges $\sigma(a, z)$, this constant charge density σ_+ does not depend on the longitudinal component z.

In conclusion we may say that the total surface charge density along the wire, not taking into account the motional electric field and the induction of charges in the conductor due to external charges, is given by the constant σ_+ added to the σ given by Eq. (6.17).

In our analysis of the radial Hall effect we are not considering the motional electric field already discussed as it is not yet completely clear if it exists or not. The results of this Section are completely theoretical. They are based upon the equilibrium of a magnetic force (due to the poloidal magnetic field) and an electric force (orthogonal to the axis of the wire) acting upon a drifting electron moving along the axis of the wire. We are not aware of any experiments which tried to measure the internal density of charges ρ_{c-} in current carrying metallic conductors.

We now compare all three components of the electric field outside the wire with one another.

6.5 Discussion

The solutions presented here will remain valid in the case of a hollow cylindrical shell of internal radius a_i and external radius a. The internal density of surface charge at $\rho = a_i$ will be zero taking into account the approximations considered here, while the external density of surface charge will be the same as obtained before. The main difference is that the electric field in the region $\rho < a_i$ will not produce any current as there is no conductor in this region.

Although many authors forget about the zeroth order force F_0 due to electrostatic induction when dealing with a current-carrying wire interacting with an external charge, there is no doubt it exists. Comparing the three components of the force already discussed, it is the only one which diverges as we approach the wire. If we are far away from the wire (at a distance $\rho \gg a$ from it) this zeroth order force falls as $1/\rho^2 \ln(\rho/a)$ (as we saw in Eqs. (4.81) and (4.83)), while the radial component of the force proportional to the voltage of the battery and of the second order force, F_1 and F_2, fall as $1/\rho$ (as we saw in Eqs. (6.21) and (6.27)).

We now compare the three components of this force given by Eqs. (4.81) and (4.82), (6.21) and (6.27). To this end we consider a particular example with orders of magnitudes similar to those employed in Sansbury's experiment [178]. He utilized a U-shaped copper current conductor (50 cm long legs spaced 10 cm apart, with 0.95 cm diameter). As we will utilize his dimensions in a different configuration (straight wire instead of a U-shaped conductor), we will consider our straight wire having a total length of $\ell = 1.20$ m and a radius $a = 4.75 \times 10^{-3}$ m. The conductivity of copper is $g = 5.7 \times 10^7$ $\Omega^{-1}\text{m}^{-1}$ and it has a number of free electrons per unit volume given by $n = 8.5 \times 10^{28}$ m^{-3}. The resistance of the wire is then given by $R = \ell/g\pi a^2 = 3.0 \times 10^{-4}$ Ω. He passed a current of 900 A in his wire, which means a potential difference between the extremities of the wire as given by $\phi_L - \phi_R = 0.27$ V. The drifting velocity in this case amounts to $v_d = I/\pi a^2 en = 0.9 \times 10^{-3}$ m/s. We will suppose moreover the symmetrical case in which $\phi_R = -\phi_L = -0.135$ V. The test charge will be the one estimated by Sansbury, namely, $q \approx 5 \times 10^{-10}$ C, at a distance of $\rho = 3.5$ cm $= 3.5 \times 10^{-2}$ m from the wire. This yields $a/\rho' = 0.121$. Although his test charge was spread over a 2 cm $\times$ 2 cm silver foil, here we suppose the test charge concentrated in a point.

These values in Eqs. (4.81) and (4.82) yield $\alpha_L = 0.247$, $F_0 = 4.5 \times 10^{-7}$ N and $E_0 = F_0/q = 9.0 \times 10^2$ V/m. Although Sansbury observed a zeroth order force, he did not measure it. For comparison we present here the zeroth order force upon an electron ($q = -1.6 \times 10^{-19}$ C) and upon a typical charge generated by friction ($q \approx 10^{-6}$ C) at the same distance from the same wire, namely: $F_0 = 4.6 \times 10^{-26}$ N and $F_0 = 1.8$ N, respectively. The huge difference between these forces arises from the fact that F_0 is proportional to the square of q. The corresponding zeroth order electric fields due to the electron and to the charge generated by friction are given by $E_0 = 2.9 \times 10^{-7}$ V/m and $E_0 = 1.8 \times 10^6$ V/m, respectively.

We now consider the force F_1 and electric field $E_1 = F_1/q$ proportional to the voltage of the battery. We consider only the radial component along the $\hat{\rho}$ direction given by Eq. (6.21). This component depends upon the values of the potentials at the extremities of the wire and also upon the value of z. With the given symmetrical potentials, $\phi_L + \phi_R = 0$ V, then the radial components of F_1 and of E_1 go to zero at $z = 0$. The maximal magnitudes of F_1 and of E_1 happen at $z = \pm \ell/2$. At these locations, with $q = 5 \times 10^{-10}$ C and with the given conditions we obtain: $F_1 = 3.5 \times 10^{-10}$ N and $E_1 = 0.69$ V/m.

As regards the second order effect, we utilize Eq. (6.27). With $q = 5 \times 10^{-10}$ C and the given conditions we obtain: $F_2 = 1.2 \times 10^{-15}$ N. This yields $E_2 = F_2/q = 2.4 \times 10^{-6}$ V/m.

Finally we can compare the three force components along the radial direction (for F_1 we consider only the maximal value). Utilizing $q = 5 \times 10^{-10}$ C we obtained: $F_0 = 4.5 \times 10^{-7}$ N, $F_1 = 3.5 \times 10^{-10}$ N and $F_2 = 1.2 \times 10^{-15}$ N. The corresponding components of the electric field were given by: $E_0 = 9.0 \times 10^2$ V/m, $E_1 = 6.9 \times 10^{-1}$ V/m and $E_2 = 2.4 \times 10^{-6}$ V/m. This yields $F_0/F_1 = 1.3 \times 10^3$, $F_1/F_2 = 2.9 \times 10^5$, $E_0/E_1 = 1.3 \times 10^3$ and $E_1/E_2 = 2.9 \times 10^5$. That is, in this case $F_0 \gg F_1 \gg F_2$ and $E_0 \gg E_1 \gg E_2$.

Similar order of magnitudes are obtained in the experiment of Bartlett and Maglic [179].

To facilitate the detection of the force F_1 it would be better not to place any test charge close to the wire. Instead of this, it would be ideal to bring a small neutral conductor close to the wire. In principle it would not act upon the wire. But when we pass a current upon the resistive wire, this wire should become charged along its surface. Therefore, it should generate an electric field E_1 outside it. This electric field would then polarize the small conductor outside it. Consequently, there would arise an attraction between the conductor and the current-carrying wire. We have already seen experiments of this kind in Chapter 3.

The second possibility in order to facilitate the detection of the force F_1 even in the presence of the force F_0 (in the case in which we approach a charged body to the current-carrying wire) would be to increase the voltage of the battery connected to the wire. As F_1 is proportional to the emf of the battery, we can make F_1 greater than F_0 working with high resistance wires connected to high voltages. We also saw experiments of this kind in Chapter 3.

In many cases we will have $F_0 \gg F_1 \gg F_2$. Despite this fact the force $\vec{F}_1$ has already been observed in the laboratory, as we saw in Chapter 3. We consider the current flowing in the top part of a circuit like that of our Figure 6.1, with symmetrical potentials: $\phi_R = -\phi_L$. In order to compare these theoretical results with the experiments, we need to obtain the lines of electric field. To obtain these lines we follow the approach presented in Sommerfeld's book [208, pp. 125-130] (German original from 1948 based on lectures delivered in 1933-1934). We obtain this in the plane xz ($y = 0$). Any plane containing the z axis will yield a similar solution. The lines of electric field are orthogonal trajectories to the equipotential lines. As $\vec{E} = -\nabla\phi$, the electric field points along the direction of the maximum space rate of change of ϕ. We are then looking for a function $\xi(\rho, z)$ such that

$$\nabla\xi(\rho, z) \cdot \nabla\phi(\rho, z) = 0 \ . \tag{6.33}$$

For $\rho < a$ we have ϕ as a linear function of z, such that ξ can be found proportional to ρ. We write it as $\xi(\rho < a, z) = -A\ell\rho$, with A as a constant. The equipotential lines, $\phi(\rho, z) = $ constant, can be written as $z_1(\rho) = K_1$, where K_1 is a constant (for each constant we have a different equipotential line). Analogously, the lines of electric field will be given by $z_2(\rho) = K_2$, where K_2 is another constant (for each K_2 we have a different line of electric field). From Eq. (6.33) we get $dz_2/d\rho = -1/(dz_1/d\rho) = (\partial\phi/\partial z)/(\partial\phi/\partial\rho)$. Integrating this equation we can obtain $\xi(\rho, z)$. With Eq. (6.9) this yields the solution for $\rho > a$. We are then led to:

$$\xi(\rho, z) = (\phi_R - \phi_L)\frac{\rho}{\ell} \quad \text{if } \rho < a \ , \tag{6.34}$$

$$\xi(\rho, z) = (\phi_R + \phi_L)\frac{z}{\ell} + (\phi_R - \phi_L)\left(\frac{\rho^2}{2\ell^2} + \frac{z^2}{\ell^2} - \frac{\rho^2}{\ell^2}\ln\frac{\rho}{\ell}\right) \quad \text{if } \rho > a \ . \tag{6.35}$$

From these equations we can easily verify Eq. (6.33).

In order to compare these results with the experiments of Bergmann, Schaefer, Jefimenko, Barnett and Kelly we need essentially the value of ℓ/a. From Figure 3.1 we get $\ell/a \approx 33$, from Figure 3.3 we get $\ell/a \approx 13$, while from Figure 3.10 we get $\ell/a \approx 4$. The plots of the equipotentials between $z = -\ell/2$ and $\ell/2$ given by Eqs. (6.8) and (6.9) with these values of ℓ/a are given in Figures 6.3, 6.4 and 6.5 (with the experimental results of Bergmann, Schaefer, Jefimenko, Barnett and Kelly overlaid on them).

Plots of the lines of electric field given by Eqs. (6.34) and (6.35) with these values of ℓ/a are given in Figures 6.6, 6.7 and 6.8 (with the experimental results of Bergmann, Schaefer, Jefimenko, Barnett and Kelly overlaid on them).

These theoretical Figures overlaid on the experimental ones indicate a very good agreement between theory and experiment.

We now consider Sansbury's experiment discussed in Chapter 3. The observed force was of the order of 10^{-7} N, although he was not able to make precise measurements. He analyzed briefly the possibility that this extra force might be the force F_1 discussed here, but only considered the longitudinal electric field outside the wire. He then concluded that this force would be three orders of magnitude smaller than the effect he measured. However, he was not aware of the radial component of $\vec{E}_1$, which can be larger than the longitudinal component, as we showed here. Moreover, his U-shaped wire was bent close to the foil

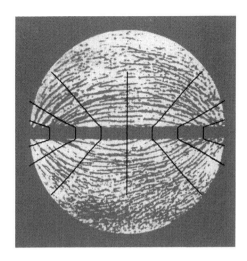

Figure 6.3: Theoretical equipotential lines overlaid on the experimental lines of electric field obtained by Bergmann and Schaefer.

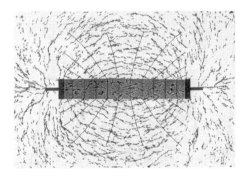

Figure 6.4: Theoretical equipotential lines overlaid on the experimental lines of electric field obtained by Jefimenko.

and thus the approximation to a long straight wire may not be applicable. Close to a corner the electric field outside the wire is even larger than the longitudinal one inside it [167]. Possibly what Sansbury detected directly was the force F_1 discussed here. It would be important to repeat his experiment carefully taking this into account.

In this Chapter we have seen a first example in which the electric field inside and outside a resistive wire carrying a steady current is due to charges spread along the surface of the conductor. The density of these surface charges is constant in time but varies along the length of the wire. It is proportional to the voltage generated by the battery connected to the wire. As the internal and external electric field is produced by charges at the surface of the wire, we can see a direct connection between electrostatics (represented by Gauss's law) and circuit theory (represented by Ohm's law). This allows a connection between these

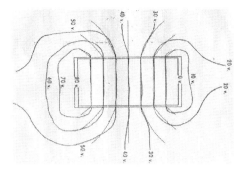

Figure 6.5: Theoretical equipotential lines overlaid on the experimental ones obtained by Jefimenko, Barnett and Kelly.

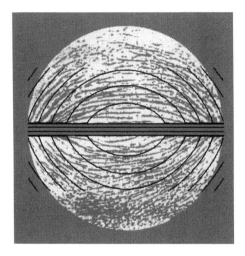

Figure 6.6: Theoretical lines of electric field overlaid on the experimental ones obtained by Bergmann and Schaefer.

two topics which are usually considered separately in the textbooks. Despite this fact some authors have called attention to the strong connection between these two branches of electromagnetism, beginning with Weber and Kirchhoff, as we see in the Appendices. Some modern scientists mention the same aspect [209, 210, 211, 171] [166, Chapter 18: A Microscopic View of Electric Circuits, pp. 623-666].

The example discussed here is important to show clearly the existence of an external electric field proportional to the potential difference acting upon the resistive wire, even in the case of a straight wire carrying a steady current. This electric field does not depend upon a variable current (with a longitudinal acceleration of the electrons along the direction of the wire), nor of a centripetal acceleration of the conduction electrons (due to any curvature in the wire). That is, this external electric field will exist even when there is no acceleration of the

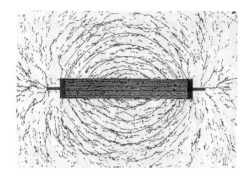

Figure 6.7: Theoretical lines of electric field overlaid on the experimental ones obtained by Jefimenko.

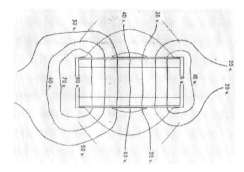

Figure 6.8: Theoretical lines of electric field overlaid on the experimental equipotential lines obtained by Jefimenko, Barnett and Kelly.

conduction electrons.

Chapter 7

Coaxial Cable

7.1 Introduction

Many authors studied the distribution of surface charges in resistive coaxial cables carrying steady currents, as well as the potential and electric field inside and outside the conductors [212, pp. 175-184] [170] [208, pp. 125-130] [213] [214] [176, pp. 318 and 509-511] [12] [215] [16, pp. 336-337] [216] [217].

Here we present the main results in this configuration considering the general case of a return conductor of finite area and finite conductivity. In this case there will be an electric field outside the external return conductor, although the magnetic field goes to zero in this region.

The configuration of the problem is that of Figure 7.1.

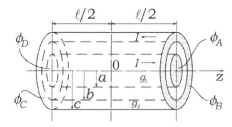

Figure 7.1: Configuration of the problem.

A constant current I flows uniformly in the z direction along the inner conductor (radius a and conductivity g_1), returning uniformly along the outer conductor (internal and external radii b and c, respectively, and conductivity g_3). The conductors have uniform circular cross-sections and a length $\ell \gg c > b > a$ centered on $z = 0$. The medium outside the conductors is considered to be air or vacuum. The potentials at the extremities located at $z = \ell/2$ of the inner and outer conductors are maintained at the constant values ϕ_A and ϕ_B, respectively. The potentials at the extremities located at $z = -\ell/2$ of the outer and inner conductors are maintained at the constant values ϕ_C and ϕ_D, respectively.

7.2 Potentials and Fields

We are interested in calculating the potentials and fields in a point $\vec{r} = (\rho, \varphi, z)$ such that $\ell \gg \rho$ and $\ell \gg |z|$, so that we can neglect edge effects. All solutions presented here were obtained with this approximation. With this approximation and configuration we then have the potential as a linear function of z. See Section 5.4. In order to have uniform currents flowing in the z direction along the inner and outer conductors, with a potential satisfying the given values at the extremities, we have:

$$\phi(\rho \leq a, \varphi, z) = \frac{\phi_A + \phi_D}{2} + (\phi_A - \phi_D)\frac{z}{\ell} , \qquad (7.1)$$

$$\phi(b \leq \rho \leq c, \varphi, z) = \frac{\phi_C + \phi_B}{2} + (\phi_B - \phi_C)\frac{z}{\ell} . \qquad (7.2)$$

By Ohm's law (with R_1 and R_3 being the resistances of the inner and outer conductors, respectively) we obtain:

$$\phi_D - \phi_A = R_1 I = \frac{\ell I}{\pi g_1 a^2} , \qquad (7.3)$$

$$\phi_B - \phi_C = R_3 I = \frac{\ell I}{\pi g_3 (c^2 - b^2)} . \qquad (7.4)$$

In the four regions ($\rho < a$, $a < \rho < b$, $b < \rho < c$ and $c < \rho$) the potential ϕ satisfies Laplace's equation $\nabla^2 \phi = 0$. By Eqs. (7.1) and (7.2) we have the value of ϕ in the first and third regions, which also supply the boundary conditions at $\rho = a$ and at $\rho = b$ in order to find ϕ in the second region. To find ϕ in the fourth region we need another boundary condition, in addition to the value of ϕ at $\rho = c$, which is given by Eq. (7.2). We then impose the following boundary condition:

$$\phi(\rho = \ell, \varphi, z) = 0 \text{ V} . \qquad (7.5)$$

This is the main non-trivial boundary condition for this problem. The same reasoning was utilized in Section 6.2 after Eq. (6.21). This equation says that the potential goes to zero at a radial distance $\rho = \ell$, so that the length ℓ of the cable appears in the solution. The usual condition $\phi(\rho \to \infty, \varphi, z) = 0$ V does not work in the situation considered here. We first tried this last condition but could not obtain a correct solution for the potential, and only discovered Eq. (7.5) working backwards. That is, from the work of Russell we knew that in general the density of the surface charges on a system of long parallel homogeneous conductors in steady-state (as is the case of the coaxial cable being considered here) varies linearly with distance along the direction of their common axis [9]. That is, if d represents a, b or c, the surface charge densities at these surfaces must be given by $\sigma_d(z) = A_d + B_d z$, with the constants A_d and B_d characterizing each surface. We then obtained the potential at all points in space by

$$\phi(\vec{r}) = \frac{1}{4\pi\varepsilon_0} \sum_{j=1}^{3} \int\int_{S_j} \frac{\sigma(\vec{r}_j) da_j}{|\vec{r} - \vec{r}_j|} . \qquad (7.6)$$

Here the sum goes over the three surfaces $\rho = a$, b and c, extending from $z = -\ell/2$ to $z = \ell/2$. After solving these integrals we discovered that ϕ went to zero not at infinity, but at $\rho = \ell$. Although this difference is important mathematically in order to arrive at a working solution, physically we can say that the potential going to zero at $\rho = \ell$ is equivalent to it going to zero at infinity. As we suppose $\ell \gg c > b > a$, we are essentially imposing that the potential goes to zero at a large distance from the cable, which is reasonable.

Here we reverse the argument, as this is more straightforward. That is, we begin with the boundary conditions for ϕ, obtaining the solutions of Laplace's equation, the electric field $\vec{E} = -\nabla\phi$ and then σ by Gauss's law.

The boundary conditions are then the values of ϕ at $\rho = a$, $\rho = b$, $\rho = c$ and $\rho = \ell$. They are given by Eqs. (7.1), (7.2) and (7.5). The solutions of Laplace's equation $\nabla^2\phi = 0$ for $a \leq \rho \leq b$ and for $c \leq \rho$ in cylindrical coordinates satisfying these boundary conditions yield:

$$\phi(a \leq \rho \leq b, \varphi, z) = \frac{\phi_B + \phi_C}{2} + (\phi_B - \phi_C)\frac{z}{\ell}$$

$$+ \left[\frac{\phi_A + \phi_D - \phi_C - \phi_B}{2} + (\phi_A - \phi_D + \phi_C - \phi_B)\frac{z}{\ell}\right]\frac{\ln(b/\rho)}{\ln(b/a)} , \qquad (7.7)$$

$$\phi(c \leq \rho, \varphi, z) = \left[\frac{\phi_C + \phi_B}{2} + (\phi_B - \phi_C)\frac{z}{\ell}\right]\frac{\ln(\ell/\rho)}{\ln(\ell/c)} . \qquad (7.8)$$

The lines of electric field are given by a function $\xi(\rho, z)$ such that $\nabla\xi \cdot \nabla\phi = 0$. By the procedure described in the previous Chapter we obtain

$$\xi(\rho < a, \varphi, z) = -(\phi_A - \phi_D)\frac{\rho}{\ell} , \qquad (7.9)$$

$$\xi(a < \rho < b, \varphi, z) = \frac{\phi_A + \phi_D - \phi_C - \phi_B}{2}\frac{z}{\ell} + \frac{\phi_B - \phi_C}{2}\frac{\rho^2}{\ell^2}\ln\frac{b}{a}$$

$$+ \frac{\phi_A - \phi_D + \phi_C - \phi_B}{2}\left(\frac{z^2}{\ell^2} + \frac{\rho^2}{2\ell^2} - \frac{\rho^2}{\ell^2}\ln\frac{\rho}{b}\right) , \qquad (7.10)$$

$$\xi(b < \rho < c, \varphi, z) = -(\phi_B - \phi_C)\frac{\rho}{\ell} , \qquad (7.11)$$

$$\xi(c < \rho, \varphi, z) = \frac{\phi_B + \phi_C}{2}\frac{z}{\ell} + \frac{\phi_B - \phi_C}{2}\left(\frac{z^2}{\ell^2} + \frac{\rho^2}{2\ell^2} - \frac{\rho^2}{\ell^2}\ln\frac{\rho}{\ell}\right) . \qquad (7.12)$$

The electric field $\vec{E} = -\nabla\phi$ is given by

$$\vec{E}(\rho < a, \varphi, z) = \frac{\phi_D - \phi_A}{\ell}\hat{z}, \tag{7.13}$$

$$\vec{E}(a < \rho < b, \varphi, z) = \left[\frac{\phi_A + \phi_D - \phi_C - \phi_B}{2}\right.$$

$$\left. + (\phi_A - \phi_D + \phi_C - \phi_B)\frac{z}{\ell}\right]\frac{1}{\ln(b/a)}\frac{\hat{\rho}}{\rho}$$

$$+ \left[\frac{\phi_C - \phi_B}{\ell} + \frac{\phi_D - \phi_A + \phi_B - \phi_C}{\ell}\frac{\ln(b/\rho)}{\ln(b/a)}\right]\hat{z}, \tag{7.14}$$

$$\vec{E}(b < \rho < c, \varphi, z) = \frac{\phi_C - \phi_B}{\ell}\hat{z}, \tag{7.15}$$

$$\vec{E}(c < \rho, \varphi, z) = \left[\frac{\phi_C + \phi_B}{2} + (\phi_B - \phi_C)\frac{z}{\ell}\right]\frac{1}{\ln(\ell/c)}\frac{\hat{\rho}}{\rho}$$

$$+ \frac{\phi_C - \phi_B}{\ell}\frac{\ln(\ell/\rho)}{\ln(\ell/c)}\hat{z}. \tag{7.16}$$

The main points to be emphasized here are the solutions (7.8) and (7.16). They show the existence of an electric field outside the resistive cable even when it is carrying a constant current.

Here we do not consider the motional electric field proportional to second order in v_d/c. Its order of magnitude is much smaller than the one considered here (proportional to the potential difference along the cable). For this reason we do not need to take it into account here.

The surface charge densities σ along the inner conductor ($\rho = a$, $\sigma_a(z)$) and along the inner and outer surfaces of the return conductor ($\rho = b$, $\sigma_b(z)$ and $\rho = c$, $\sigma_c(z)$) can be obtained easily utilizing Gauss's law:

$$\oint_S \vec{E} \cdot d\vec{a} = \frac{Q}{\varepsilon_0}, \tag{7.17}$$

where $d\vec{a}$ is the surface element pointing normally outwards the closed surface S and Q is the net charge inside S. This yields $\sigma_a(z) = \varepsilon_0 E_{2\rho}(\rho \to a, z)$, $\sigma_b(z) = -\varepsilon_0 E_{2\rho}(\rho \to b, z)$ and $\sigma_c(z) = \varepsilon_0 E_{4\rho}(\rho \to c, z)$, where the subscripts 2ρ and 4ρ mean the radial component of $\vec{E}$ in the second and fourth regions, $a < \rho < b$ and $c < \rho$, respectively. This means that:

$$\sigma_a(z) = \frac{\varepsilon_0}{a}\frac{1}{\ln(b/a)}\left[\frac{\phi_A + \phi_D - \phi_C - \phi_B}{2} + (\phi_A - \phi_D + \phi_C - \phi_B)\frac{z}{\ell}\right], \tag{7.18}$$

$$\sigma_b(z) = -\frac{a}{b}\sigma_a(z) \;, \tag{7.19}$$

$$\sigma_c(z) = \frac{\varepsilon_0}{c}\frac{1}{\ln(\ell/c)}\left[\frac{\phi_C + \phi_B}{2} + (\phi_B - \phi_C)\frac{z}{\ell}\right] \;. \tag{7.20}$$

An alternative way of obtaining ϕ and $\vec{E}$ is to begin with the surface charges as given by Eqs. (7.18) to (7.20). We then calculate the electric potential ϕ (and $\vec{E} = -\nabla\phi$) through Eq. (7.6). We checked the calculations with this procedure.

7.3 The Symmetrical Case

In order to visualize the equipotentials and lines of electric field we consider $\ell/c = 5$, $\ell/b = 15/2$ and $\ell/a = 15$. There are two main cases of interest, the symmetrical and asymmetrical cases. In the symmetrical case there are two equal batteries located at both extremities of the cable, Figure 7.2.

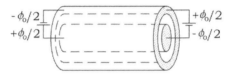

Figure 7.2: The symmetrical case.

They generate potentials $\phi_B = \phi_D = -\phi_A = -\phi_C \equiv \phi_0/2$. In this case the surface charge densities go to zero at the center of the cable ($z = 0$) in all three surfaces ($\rho = a$, b and c). The equipotentials and lines of electric field for this situation are shown in Figures 7.3 and 7.4, respectively.

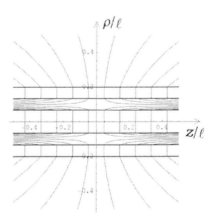

Figure 7.3: Equipotential lines for the symmetrical case.

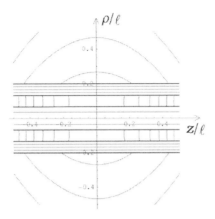

Figure 7.4: Lines of electric field for the symmetrical case.

In this case the potential is simply proportional to z without any additive constant. We can then write it in terms of the currents and conductivities as given by:

$$\phi(\rho \leq a) = -\frac{Iz}{\pi g_1 a^2}, \tag{7.21}$$

$$\phi(a \leq \rho \leq b) = -\frac{I}{\pi} \frac{z}{\ln(b/a)} \left[\frac{\ln(b/\rho)}{g_1 a^2} - \frac{\ln(\rho/a)}{g_3(c^2 - b^2)} \right], \tag{7.22}$$

$$\phi(b \leq \rho \leq c) = \frac{Iz}{\pi g_3(c^2 - b^2)}, \tag{7.23}$$

$$\phi(c \leq \rho) = \frac{I}{\pi} \frac{\ln(\ell/\rho)}{\ln(\ell/c)} \frac{z}{g_3(c^2 - b^2)}. \tag{7.24}$$

Particular cases include an equipotential outer conductor ($\phi_C = \phi_B = 0$) with an infinite area ($c \to \infty$) or with an infinite conductivity ($g_3 \to \infty$). These solutions are recovered taking $g_3(c^2 - b^2) \to \infty$, such that $\sigma_c(z) \to 0$, $\vec{E}(\rho > b) \to 0$ and $\phi(\rho \geq b) \to 0$ for any z. The opposite solution when the current flows in an inner conductor of infinite conductivity, returning in an outer conductor of finite area and finite conductivity, is also easily obtained from the previous result, yielding $\vec{E}(\rho < a) \to 0$ and $\phi(\rho \leq a) \to 0$ for any z.

7.4 The Asymmetrical Case

In the asymmetrical case there is a battery at the left extremity and a load resistance R_L at the right extremity, Figure 7.5.

We can represent the potentials generated by the battery producing a voltage ϕ_0 between its terminals as $\phi_D = -\phi_C \equiv \phi_0/2$. By Ohm's law the total current

Figure 7.5: The asymmetrical case.

I is related to the total resistance $R_t \equiv R_1 + R_L + R_2$ by $I = \phi_0/R_t$. Analogously: $\phi_D - \phi_A = \phi_0(R_1/R_t)$, $\phi_A - \phi_B = \phi_0(R_L/R_t)$ and $\phi_B - \phi_C = \phi_0(R_2/R_t)$. These results in Eqs. (7.1) to (7.12) yield:

$$\phi(\rho \leq a, \varphi, z) = \phi_0 \left(\frac{R_2 + R_L}{2R_t} - \frac{R_1}{R_t} \frac{z}{\ell} \right), \tag{7.25}$$

$$\phi(a \leq \rho \leq b, \varphi, z) = -\phi_0 \left[\left(\frac{R_1 + R_L}{2R_t} - \frac{R_1 + R_2 + 2R_L}{2R_t} \frac{\ln(b/\rho)}{\ln(b/a)} \right) \right.$$
$$\left. - \left(\frac{R_2}{R_t} - \frac{R_1 + R_2}{R_t} \frac{\ln(b/\rho)}{\ln(b/a)} \right) \frac{z}{\ell} \right], \tag{7.26}$$

$$\phi(b \leq \rho \leq c, \varphi, z) = -\phi_0 \left(\frac{R_1 + R_L}{2R_t} - \frac{R_2}{R_t} \frac{z}{\ell} \right), \tag{7.27}$$

$$\phi(c \leq \rho, \varphi, z) = -\phi_0 \left(\frac{R_1 + R_L}{2R_t} - \frac{R_2}{R_t} \frac{z}{\ell} \right) \frac{\ln(\ell/\rho)}{\ln(\ell/c)}. \tag{7.28}$$

$$\xi(\rho < a, \varphi, z) = \phi_0 \frac{R_1}{R_t} \frac{\rho}{\ell}, \tag{7.29}$$

$$\xi(a < \rho < b, \varphi, z) = \phi_0 \left[\frac{R_1 + R_2 + 2R_L}{2R_t} \frac{z}{\ell} + \frac{R_2}{2R_t} \frac{\rho^2}{\ell^2} \ln \frac{b}{a} \right.$$
$$\left. - \frac{1}{2} \left(\frac{z^2}{\ell^2} + \frac{\rho^2}{2\ell^2} - \frac{\rho^2}{\ell^2} \ln \frac{\rho}{b} \right) \right], \tag{7.30}$$

$$\xi(b < \rho < c, \varphi, z) = -\phi_0 \frac{R_2}{R_t} \frac{\rho}{\ell}, \tag{7.31}$$

$$\xi(c < \rho, \varphi, z) = -\phi_0 \left[\frac{R_1 + R_L}{2R_t} \frac{z}{\ell} - \frac{R_2}{2R_t} \left(\frac{z^2}{\ell^2} + \frac{\rho^2}{2\ell^2} - \frac{\rho^2}{\ell^2} \ln \frac{\rho}{\ell} \right) \right]. \tag{7.32}$$

These Equations are plotted in Figures 7.6 and 7.7 when $\ell/c = 5$, $\ell/b = 15/2$, $\ell/a = 15$ and $R_1 = R_2 = R_L$.

As we obtained algebraic solutions for the fields, potentials and surface charges, it is easy to apply them for commercial cables. In this way we can know the orders of magnitude of these quantities for several standard cables.

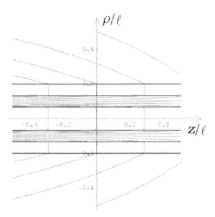

Figure 7.6: Equipotential lines for the asymmetrical case.

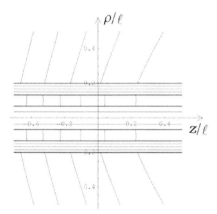

Figure 7.7: Lines of electric field for the asymmetrical case.

7.5 Discussion

The distribution of charges given by Eqs. (7.18) to (7.20) is equivalent to equal and opposite charges in the facing surfaces. That is, the charge at the position $\rho = a$, z, in a length dz, $dq_a(z) = 2\pi a \sigma_a(z) dz$, is equal and opposite to the charge at the position $\rho = b$, z, in the same length dz: $dq_b(z) = 2\pi b \sigma_b(z) dz = -dq_a(z)$.

The electric field outside the coaxial cable then depends only on the surface charges at the external wall of the return conductor, $\sigma_c(z)$:

$$\phi(c \leq \rho, \varphi, z) = \frac{c}{\varepsilon_0} \sigma_c(z) \ln \frac{\ell}{\rho} = \left[\frac{\phi_B + \phi_C}{2} + (\phi_B - \phi_C) \frac{z}{\ell} \right] \frac{\ln(\rho/\ell)}{\ln(c/\ell)} . \quad (7.33)$$

The main nontrivial conclusions of this analysis are Eqs. (7.16) and (7.33). They show that although there is no vector potential or magnetic field outside a coaxial cable, the electric field will be different from zero when there is a finite resistivity in the outer conductor. To our knowledge the first to mention this external electric field outside a resistive coaxial cable was Russell in his important paper of 1983 [213]. The solution of this Chapter presents an analytical calculation of this field.

This external electric field indicates that there is no shielding in a coaxial cable with a resistive outer conductor (sheath). It is important to realize this specially when dealing with interferences in telecommunication systems. Even with a long cable there will be this external electric field, as can be seen from Eq. (7.33). For this reason this resistive cable will influence other electrical systems nearby. This field will be present even for variable current. This is a relevant aspect neglected by most authors.

Chapter 8

Transmission Line

8.1 Introduction

One of the most important electrical systems is that of a two-wire transmission line, usually called twin-leads. We consider here homogeneous resistive wires fixed in the laboratory and carrying steady currents. The goal here is to calculate the electric field outside the wires.

The case of twin-leads was first considered by Stratton [218, p. 262]. Although he called attention to the electric field outside the transmission line, this has been forgotten by most authors, as we have seen. We treated this case in more detail in 1999 [219] and here we follow this latter approach. These are the only theoretical works dealing with this configuration known to us.

8.2 Two-Wire Transmission Line

The configuration of the system is given in Figure 8.1.

We have two equal straight wires of circular cross-sections of radii a and length ℓ, surrounded by air. Their axes are separated by a distance b and are parallel to the z axis, symmetrically located relative to the z and x axes. That is, the centers of the wires are located at $(x, y, z) = (-b/2, 0, 0)$ and $(+b/2, 0, 0)$. The conductivity of the wires is g and their extremities are located at $z = -\ell/2$ and $z = +\ell/2$. Here we calculate the electric potential ϕ and the electric field $\vec{E}$ at a point (x, y, z) such that $\ell \gg r = \sqrt{x^2 + y^2 + z^2}$. Moreover, we also assume that $\ell \gg b/2 > a$, so that we can neglect edge effects.

We want to find the potential and electric field when a current I flows uniformly through one of the wires along the direction $+\hat{z}$ and returns uniformly through the other wire along the direction $-\hat{z}$. The current densities in both wires are then given by $\vec{J} = (I/\pi a^2)\hat{z}$ and $\vec{J} = -(I/\pi a^2)\hat{z}$, respectively. As we are considering homogeneous wires with a constant conductivity g, Ohm's law yields the internal electric field in the wires as $\vec{E} = \pm(I/g\pi a^2)\hat{z}$. We do not need to consider in $\vec{E}$ the influence of the time variation of the vector potential

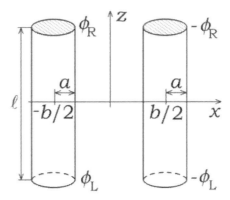

Figure 8.1: Two homogeneously resistive parallel wires of radii a separated by a distance b. The first wire carries a steady current I along the positive z direction while the second wire carries the return current I along the negative z direction.

as we are dealing with a steady current in stationary wires, so that $\partial \vec{A}/\partial t = \vec{0}$ everywhere. We can then write $\vec{E} = -\nabla \phi$. As we have a constant electric field in each wire, this implies that the potential is constant over each cross-section and a linear function of z. In this work we consider a symmetrical situation for the potentials so that in the first wire the current flows from the potential ϕ_L at $z = -\ell/2$ to ϕ_R at $z = \ell/2$ and returns in the second wire from $-\phi_R$ at $z = \ell/2$ to $-\phi_L$ at $z = -\ell/2$, Figure 8.1. We can then write:

$$\phi_F(z) = \frac{\phi_R + \phi_L}{2} + (\phi_R - \phi_L)\frac{z}{\ell} = \frac{\phi_R + \phi_L}{2} + \frac{I}{g\pi a^2}z \,, \tag{8.1}$$

$$\phi_S(z) = -\phi_F(z) \,. \tag{8.2}$$

In these equations $\phi_F(z)$ and $\phi_S(z)$ are the potentials as a function of z over the cross-section of the first and second conductors, respectively.

In this Chapter we neglect the small Hall effect due to the azimuthal magnetic field generated by these currents. See Section 6.4. This effect creates a redistribution of the charge density within the wires, and modifies the surface charges also. As these are usually small effects, they will not be considered here.

We now find the potential in space supposing there is air outside the conductors. As the conductors are straight and the boundary conditions (the potentials over the surface of the conductors) are linear functions of z, the same must be valid everywhere, as we saw in Section 5.4. That is, $\phi = (A + Bz)f(x, y)$, where A and B are constants and $f(x, y)$ is a function of x and y. This function can be found by the method of images, imposing a constant potential ϕ_0 over the first wire and $-\phi_0$ over the second one [13, Section 2.1]. The final solution for ϕ and $\vec{E}$ satisfying the given boundary conditions, valid for the region outside the wires, is given by:

$$\phi(x,\ y,\ z) = -\left(\frac{\phi_R + \phi_L}{2} + (\phi_R - \phi_L)\frac{z}{\ell}\right)\frac{1}{2\ln\frac{b-\sqrt{b^2-4a^2}}{2a}}$$

$$\times \ln \frac{(x-\sqrt{b^2-4a^2}/2)^2 + y^2}{(x+\sqrt{b^2-4a^2}/2)^2 + y^2}\ , \qquad (8.3)$$

$$\vec{E} = -\left(\frac{\phi_R + \phi_L}{2} + (\phi_R - \phi_L)\frac{z}{\ell}\right)\frac{\sqrt{b^2-4a^2}}{\ln\frac{b+\sqrt{b^2-4a^2}}{2a}}$$

$$\times \frac{(x^2-y^2+a^2-b^2/4)\hat{x} + 2xy\hat{y}}{D_1^4}$$

$$+ \frac{\phi_R - \phi_L}{\ell}\frac{1}{2\ln\frac{b-\sqrt{b^2-4a^2}}{2a}}\left[\ln\frac{(x-\sqrt{b^2-4a^2}/2)^2+y^2}{(x+\sqrt{b^2-4a^2}/2)^2+y^2}\right]\hat{z}\ , \qquad (8.4)$$

where:

$$D_1^4 \equiv x^4 + y^4 + b^4/16 + a^4 + 2x^2y^2 - b^2x^2/2$$

$$+ 2a^2x^2 + b^2y^2/2 - 2a^2y^2 - b^2a^2/2\ . \qquad (8.5)$$

The equipotentials at $z = 0$ are plotted in Figure 8.2.

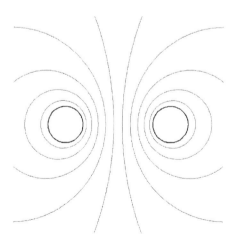

Figure 8.2: Equipotentials in the plane $z = 0$.

It is also relevant to express these results in cylindrical coordinates $(\rho,\ \varphi,\ z)$ centered on the first and second wires. See Figure 8.3.

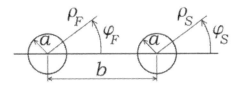

Figure 8.3: Cylindrical coordinates centered on the first and second wires.

For the first wire this can be accomplished replacing x by $\rho_F \cos\varphi_F - b/2$, y by $\rho_F \sin\varphi_F$, $\hat{x}$ by $\hat{\rho}_F \cos\varphi_F - \hat{\varphi}_F \sin\varphi_F$ and $\hat{y}$ by $\hat{\rho}_F \sin\varphi_F + \hat{\varphi}_F \cos\varphi_F$, yielding:

$$\phi(\rho_F, \varphi_F, z) = -\left(\frac{\phi_R + \phi_L}{2} + (\phi_R - \phi_L)\frac{z}{\ell}\right) \frac{1}{2\ln\frac{b-\sqrt{b^2-4a^2}}{2a}} \ln\sqrt{\frac{D_2^2}{D_3^2}}, \quad (8.6)$$

where:

$$D_2^2 \equiv \rho_F^2 - \rho_F(\cos\varphi_F)\left(b + \sqrt{b^2 - 4a^2}\right) + \frac{b^2}{2} - a^2 + \frac{b\sqrt{b^2-4a^2}}{2}, \quad (8.7)$$

and

$$D_3^2 \equiv \rho_F^2 - \rho_F(\cos\varphi_F)\left(b - \sqrt{b^2 - 4a^2}\right) + \frac{b^2}{2} - a^2 - \frac{b\sqrt{b^2-4a^2}}{2}. \quad (8.8)$$

The electric field is then given by:

$$\vec{E} = -\left(\frac{\phi_R + \phi_L}{2} + (\phi_R - \phi_L)\frac{z}{\ell}\right) \frac{\sqrt{b^2-4a^2}}{\ln\frac{b+\sqrt{b^2-4a^2}}{2a}}$$

$$\times \frac{(\rho_F^2 \cos\varphi_F - \rho_F b + a^2 \cos\varphi_F)\hat{\rho}_F + (\sin\varphi_F)(\rho_F^2 - a^2)\hat{\varphi}_F}{D_4^4}$$

$$+ \frac{\phi_R - \phi_L}{\ell} \frac{1}{2\ln\frac{b-\sqrt{b^2-4a^2}}{2a}}$$

$$\times \left[\ln\frac{\rho_F^2 - \rho_F(\cos\varphi_F)(b+\sqrt{b^2-4a^2}) + b^2/2 - a^2 + b\sqrt{b^2-4a^2}/2}{\rho_F^2 - \rho_F(\cos\varphi_F)(b-\sqrt{b^2-4a^2}) + b^2/2 - a^2 - b\sqrt{b^2-4a^2}/2}\right]\hat{z}, \quad (8.9)$$

where:

$$D_4^4 \equiv \rho_F^4 - 2\rho_F^3 b \cos\varphi_F + \rho_F^2 b^2 + a^4$$

$$+ 2\rho_F^2 a^2 (\cos^2\varphi_F - \sin^2\varphi_F) - 2\rho_F b a^2 \cos\varphi_F. \tag{8.10}$$

The density of surface charges over the first and second wires, σ_F and σ_S, can then be found by ε_0 times the radial component of the electric field over the surface of each cylinder, yielding:

$$\sigma_F = \left(\frac{\phi_R + \phi_L}{2} + (\phi_R - \phi_L)\frac{z}{\ell} \right) \frac{\varepsilon_0}{2a \ln \frac{b+\sqrt{b^2-4a^2}}{2a}} \frac{\sqrt{b^2-4a^2}}{b/2 - a\cos\varphi_F}, \tag{8.11}$$

$$\sigma_S = -\left(\frac{\phi_R + \phi_L}{2} + (\phi_R - \phi_L)\frac{z}{\ell} \right) \frac{\varepsilon_0}{2a \ln \frac{b+\sqrt{b^2-4a^2}}{2a}} \frac{\sqrt{b^2-4a^2}}{b/2 + a\cos\varphi_S}. \tag{8.12}$$

In order to check these results we calculated the potential ϕ inside each wire and in space, beginning with these surface charge densities and utilizing:

$$\phi(x, y, z) = \frac{1}{4\pi\varepsilon_0} \left[\int_{z'=-\ell/2}^{\ell/2} \int_{\varphi_F'=0}^{2\pi} \frac{\sigma_F(\varphi_F') a d\varphi_F' dz'}{|\vec{r} - \vec{r}\,'|} \right.$$

$$\left. + \int_{z'=-\ell/2}^{\ell/2} \int_{\varphi_S'=0}^{2\pi} \frac{\sigma_S(\varphi_S') a d\varphi_S' dz'}{|\vec{r} - \vec{r}\,'|} \right]. \tag{8.13}$$

Here we integrate over the surfaces of the first and second cylinders, S_L and S_R, respectively. We can then check these results assuming the correctness of the method of images for the electrostatic problem and utilizing the approximations $\ell \gg |\vec{r}|$ and $\ell \gg b/2 > a$.

With $b \gg a$ and $b \gg \rho_F$, Eqs. (8.11) and (8.6) yield:

$$\sigma_F \approx \frac{\varepsilon_0}{a \ln(b/a)} \left(\frac{\phi_R + \phi_L}{2} + (\phi_R - \phi_L)\frac{z}{\ell} \right), \tag{8.14}$$

and

$$\phi(a < \rho_F \ll b, \varphi_F, z) \approx \frac{a\sigma_F(z)}{\varepsilon_0} \ln \frac{b}{\rho_F}. \tag{8.15}$$

These results are analogous to Eqs. (6.2) and (6.9).

8.3 Discussion

The first aspect to be discussed here is the qualitative interpretation of these results. In all this Section we will assume $\phi_R = 0$ in order to simplify the analysis. The distribution of surface charges for a given z is similar to the distribution of charges in the electrostatic problem given the potentials ϕ_0 and $-\phi_0$ at the first and second wires, without current. That is, $\sigma_F(\varphi_F) > 0$ for any φ_F and its maximum value is at $\varphi_F = 0$ rad. The density of surface charges at the second wire, σ_S, has the same behaviour of σ_F with an overall change of sign, with its maximum magnitude occurring at $\varphi_S = \pi$ rad. A qualitative plot of the surface charges at $z = 0$ is given in Figure 8.4.

Figure 8.4: Qualitative distribution of surface charges for two parallel wires in the plane $z = 0$.

A quantitative plot of σ_F is given in Figure 8.5 supposing $b/2a = 10/3$ and normalizing the surface charge density by the value of σ_F at $\varphi_F = \pi$ rad.

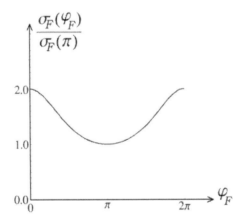

Figure 8.5: Quantitative normalized distribution of the density of surface charges in the first wire in the plane $z = 0$ as a function of the azimuthal angle.

It should also be remarked that for a fixed φ_F the surface density decreases linearly from $z = -\ell/2$ to $z = \ell/2$, the opposite happening with σ_S for a fixed φ_S.

We can integrate the surface charges over the circumference of each wire, obtaining the integrated charge per unit length $\lambda(z)$ as:

$$\lambda_F(z) = \int_{\varphi_F=0}^{2\pi} a\sigma_F(\varphi_F)d\varphi_F$$

$$= -\frac{2\pi\varepsilon_0}{\ln\left[(b-\sqrt{b^2-4a^2})/2a\right]}\left[\frac{\phi_R+\phi_L}{2} + (\phi_R-\phi_L)\frac{z}{\ell}\right]. \quad (8.16)$$

$$\lambda_S(z) = -\int_{\varphi_S=0}^{2\pi} a\sigma_S(\varphi_S)d\varphi_S = -\lambda_F(z). \quad (8.17)$$

One important aspect to discuss is the experimental relevance of these surface charges in terms of forces. That is, as the wires have a net charge in each section, there will be an electrostatic force acting on them. We can then compare this force with the magnetic force. The latter is given essentially by (force per unit length):

$$\frac{dF_M}{dz} = \frac{\mu_0 I^2}{2\pi b}, \quad (8.18)$$

where we are supposing $b/2 \gg a$.

We now calculate the electric force per unit length on the first wire, integrating the force over its circumference. We consider a typical region in the middle of the wire, around $z=0$, and once more suppose $b/2 \gg a$:

$$\frac{d\vec{F}_E}{dz} = \int_{\varphi_F=0}^{2\pi} a\sigma_F(\varphi_F)\vec{E}(\rho_F=a,\varphi_F,z=0)d\varphi_F \approx \frac{\pi\varepsilon_0\phi_L^2}{\ln^2 b/a}\left(\frac{\hat{x}}{b} + \frac{\hat{z}}{\ell}\right). \quad (8.19)$$

From Eqs. (8.18) and (8.19) the ratio of the magnetic to the radial electric force is given by (with Ohm's law $\phi_L^2/I^2 = R^2 = (\ell/g\pi a^2)^2$, R being the resistance of each wire):

$$\frac{F_M}{F_E} \approx \frac{\mu_0/\varepsilon_0}{2R^2}\ln^2\frac{b}{a}. \quad (8.20)$$

As $\mu_0/\varepsilon_0 = 1.4 \times 10^5\ \Omega^2$ this ratio will be usually many orders of magnitude greater than 1. This would be of the order of 1 when $R \approx 370\ \Omega$ (supposing $\ln(b/a) \approx 1$). This is a very large resistance for homogeneous wires.

In order to compare this force with the magnetic force we suppose typical copper wires of conductivities $g = 5.7 \times 10^7$ m$^{-1}\Omega^{-1}$, lengths $\ell = 1$ m, separated by a distance $b = 6$ mm and diameters $2a = 1$ mm. This means that by Ohm's law $\phi_L^2/I^2 = R^2 \approx 5 \times 10^{-4}\ \Omega^2$. With these values the ratio of the longitudinal electric force to the magnetic force is of the order of 7×10^{-11}, while the ratio of the radial electric force to the magnetic force is of the order of 1×10^{-8}. That is, the electric force between the wires due to these surface charges is typically 10^{-8} times smaller than the magnetic force. This shows that we can usually neglect these electric forces.

Despite this fact it should be remarked that while the magnetic force is repulsive in this situation (parallel wires carrying currents in opposite directions), the radial electric force is attractive, as we can see from the charges in Figure 8.4.

The situation described in this Chapter is very similar to the experiments performed by Bergmann, Schaefer and Jefimenko, whose results are presented in Figures 3.2 and 3.5. We can compare these experiments with the theoretical calculations by plotting the equipotentials obtained here. We need essentially the values of ℓ/b, $b/2a$ and $\ell/2a$. From Fig. 3.2 we obtain $\ell/b \approx 2.8$, $b/2a \approx 7.4$ and $\ell/2a \approx 20.7$. From Fig. 3.5 we have $\ell/b \approx 1.9$, $b/2a \approx 3.0$ and $\ell/2a \approx 5.7$. These values together with $\phi_A = 0$ V and $\phi_B = 1$ V yielded the equipotentials given by Eq. (8.3) at $y = 0$, Figures 8.6 and 8.7.

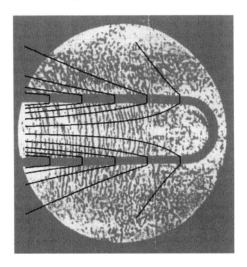

Figure 8.6: Theoretical equipotential lines overlaid on the experimental lines of electric field obtained by Bergmann and Schaefer.

The lines of electric field orthogonal to the equipotentials can be obtained by the procedure described in Sommerfeld's book, discussed in Section 6.5. This yields the following solutions in the plane $y = 0$ outside the wires:

$$\xi^{\text{out}}(x, 0, z) = -(\phi_R + \phi_L)\frac{z}{\ell} + (\phi_R - \phi_L)\left[\frac{x(x^2 - 3x_o^2)}{6x_o\ell^2}\ln\frac{(x - x_o)^2}{(x + x_o)^2}\right.$$

$$\left. + \frac{x_o^2}{3\ell^2}\ln\frac{(x - x_o)^2(x + x_o)^2}{x_o^4} - \frac{x^2}{3\ell^2} - \frac{z^2}{\ell^2}\right], \quad (8.21)$$

where $x_o \equiv \sqrt{b^2 - 4a^2}/2$.

The lines of electric field inside the first and second wires can be written as, respectively:

$$\xi^F(x, 0, z) = -(\phi_R - \phi_L)\frac{|x + b/2|}{\ell}, \quad (8.22)$$

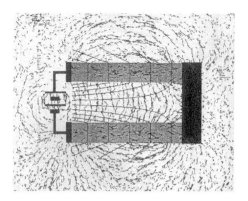

Figure 8.7: Theoretical equipotential lines overlaid on the experimental lines of electric field obtained by Jefimenko.

$$\xi^S(x, 0, z) = (\phi_R - \phi_L)\frac{|x - b/2|}{\ell} \ . \tag{8.23}$$

With the previous values of ℓ/b, $b/2a$ and $\ell/2a$ for the two experiments already mentioned we obtain the lines of electric field by these equations as given in Figures 8.8 and 8.9 (with Figure 3.2 and the left side of Figure 3.5 overlaid on them).

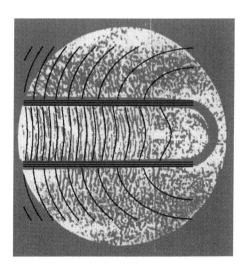

Figure 8.8: Theoretical lines of electric field overlaid on the experimental lines obtained by Bergmann and Schaefer.

These numerical plots are very similar to the experiments, especially in the region between the wires. Although this calculation is strictly valid only for $r \ll \ell$, the numerical plots go from $z = -\ell/2$ to $\ell/2$. As the result is in rea-

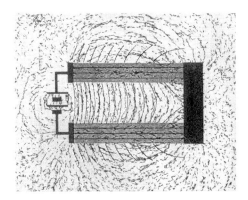

Figure 8.9: Theoretical lines of electric field overlaid on the experimental lines obtained by Jefimenko.

sonable agreement with the experiments, we conclude that the exact boundary conditions at $z = \pm \ell/2$ are not very relevant in these particular configurations.

We can also estimate the ratio of the radial component of the electric field to the axial component just outside the wire. We consider the first wire at three different values of z: $z = -\ell/2$, $z = 0$ and $z = \ell/2$. The axial component E_z is constant over the cross-section and does not depend on z. On the other hand, the radial component E_x is a linear function of z and also depends on φ_F. In this comparison we consider $\varphi_F = 0$. With these values and Jefimenko's data in Eq. (8.4) we obtain $E_x/E_z \approx 12$ at $z = -\ell/2$, 6 at $z = 0$ and 0 at $z = \ell/2$. That is, the radial component of the electric field just outside the wire is typically one order of magnitude larger than the axial electric field responsible for the current. Jefimenko's experiment gives a clear confirmation of this fact.

Chapter 9

Resistive Plates

9.1 Introduction

In this Chapter we consider one or more resistive plates carrying steady currents. We consider an ideal case of an infinite resistive bidimensional plate (like an infinite plane). The current is supposed to flow uniformly over the plate along a straight direction.

When there is no current flowing in the conducting plate and we approximate a test charge, waiting until electrostatic equilibrium is reached, with the test charge at a distance z from the plate, there will be an attraction between the plate and the charge given by Eq. (4.1).

What happens when we now pass a constant current through the stationary resistive plate connected to a battery? The electric field that maintains the current against Ohmic resistance is generated by a surface charge distribution on the plate. Our goal is to calculate the potential and electric field over the plate and in the space surrounding it when the plate carries a steady current.

The subject of this Chapter was first discussed by Jefimenko [176, pp. 303-304], and later by other authors [220, 221].

9.2 Single Plate

We consider the case of conducting plates from the point of view of surface charge distributions generating the electric fields.

The configuration we are considering is that of a rectangular plate of length ℓ_y in the y direction and ℓ_z in the z direction. The plate is located in the $x = 0$ plane with its center at $(x, y, z) = (0, 0, 0)$. We assume that the current I flows uniformly from $-\ell_z/2$ to $+\ell_z/2$ with a surface current density $\vec{K} = (I/\ell_y)\hat{z}$, Figure 9.1. We also assume that the surface charge density is linear along z, as we saw in Section 5.4:

$$\sigma(z) = \sigma_A + \sigma_B \frac{z}{\ell_z} \ . \tag{9.1}$$

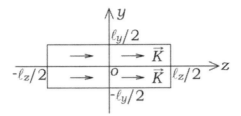

Figure 9.1: A resistive plate in the plane $x = 0$ with a steady and uniform surface current density $\vec{K}$ along the positive z direction.

Note that the surface charge should in general be a function of the y and z coordinates, $\sigma = \sigma(y, z)$. We neglect the dependence on y as an approximation for $\ell_y \gg |\vec{r}|$, where $|\vec{r}| = \sqrt{x^2 + y^2 + z^2}$ is the distance from the observation point to the center of the plate. Moreover, we consider that the test charge is far from the battery. The case in which the test charge or the observation point is close to the battery, in analogy with the case of a test charge close to the middle point of side AD of Figure 6.2, was considered in 2005 [222].

In order to generate such steady and uniform (that is, independent of the variable y) longitudinal current along an infinite plate, the ideal battery driving this current can be thought as an infinite straight line along the plate and orthogonal to the direction of the current.

The electric potential is readily given from the surface charge $\sigma(z)$ by:

$$\phi(\vec{r}) = \frac{1}{4\pi\varepsilon_0} \int\int \frac{\sigma(z')da'}{|\vec{r} - \vec{r'}|} . \qquad (9.2)$$

This integral should be evaluated over the whole charge distribution. We are interested in the potential at the symmetric plane $y = 0$:

$$\phi(x, 0, z) = \frac{1}{4\pi\varepsilon_0} \int_{-\ell_y/2}^{\ell_y/2} \int_{-\ell_z/2}^{\ell_z/2} \frac{\sigma_A + \sigma_B z'/\ell_z}{\sqrt{x^2 + y'^2 + (z - z')^2}} dy' dz' . \qquad (9.3)$$

We solve these integrals utilizing three different approximations:

$$(A) \quad \ell_y \gg \ell_z \gg \sqrt{x^2 + z^2} , \qquad (9.4)$$

$$(B) \quad \ell \equiv \ell_y = \ell_z \gg \sqrt{x^2 + z^2} , \qquad (9.5)$$

$$(C) \quad \ell_z \gg \ell_y \gg \sqrt{x^2 + z^2} . \qquad (9.6)$$

For each case the potential is given by, respectively:

$$\phi(\ell_y \gg \ell_z) \approx \frac{\sigma(z)}{2\varepsilon_0} \left(\frac{\ell_z}{\pi} - |x| \right) + \frac{\sigma_A}{2\varepsilon_0} \frac{\ell_z}{\pi} \ln \frac{2\ell_y}{\ell_z} , \qquad (9.7)$$

$$\phi(\ell_y = \ell_z \equiv \ell) \approx \frac{\sigma(z)}{2\varepsilon_0}\left(\frac{2\ell}{\pi}\ln(\sqrt{2}+1) - |x|\right) + \frac{\sigma_A}{2\varepsilon_0}\frac{\ell}{\pi}\ln(\sqrt{2}+1), \qquad (9.8)$$

$$\phi(\ell_z \gg \ell_y) \approx \frac{\sigma(z)}{2\varepsilon_0}\left(\frac{\ell_y}{\pi}\ln\frac{2\ell_z}{\ell_y} - |x|\right) + \frac{\sigma_A}{2\varepsilon_0}\frac{\ell_y}{\pi}. \qquad (9.9)$$

For each approximation we define the constants λ_1 and λ_2 by the expressions:

$$(A) \quad \lambda_1 \equiv \frac{\ell_z}{2\pi}, \quad \lambda_2 \equiv \frac{\ell_z}{2\pi}\ln\frac{2\ell_y}{\ell_z} \gg \lambda_1, \qquad (9.10)$$

$$(B) \quad \lambda_1 \equiv \frac{\ell}{2\pi}\ln(\sqrt{2}+1), \quad \lambda_2 \equiv \frac{\ell}{2\pi}\ln(\sqrt{2}+1) = \lambda_1, \qquad (9.11)$$

$$(C) \quad \lambda_1 \equiv \frac{\ell_y}{2\pi}\ln\frac{2\ell_z}{\ell_y}, \quad \lambda_2 \equiv \frac{\ell_y}{2\pi} \ll \lambda_1. \qquad (9.12)$$

The constants λ_1 and λ_2 have dimensions of length, are typically of the order of magnitude of the width or length of the plates, and are much larger than the distance to the point of interest $r = \sqrt{x^2 + z^2}$.

With these constants we can write the electric potential for this single plate in the three given cases (A), (B) and (C) as:

$$\phi(x, 0, z) = \frac{1}{\varepsilon_0}\left[\left(\sigma_A + \sigma_B\frac{z}{\ell_z}\right)\left(\lambda_1 - \frac{|x|}{2}\right) + \sigma_A \lambda_2\right]. \qquad (9.13)$$

The electric field $\vec{E} = -\nabla\phi$ is given by:

$$\vec{E}(x, 0, z) = \pm\frac{1}{\varepsilon_0}\left[\frac{\sigma_A + \sigma_B z/\ell_z}{2}\hat{x} \mp \sigma_B\frac{\lambda_1 - |x|/2}{\ell_z}\hat{z}\right], \qquad (9.14)$$

where the top (bottom) sign is for $x > 0$ ($x < 0$).

In order to test the coherence of this procedure we invert the argument. Applying Gauss's law to a small cylinder centered on the plate we obtain the usual boundary condition relating the normal component of the electric field, E_x, to the surface charge density, σ, namely: $\varepsilon_0 E_x(\lim x \to 0^+) - \varepsilon_0 E_x(\lim x \to 0^-) = \sigma(z)$. And this yields exactly the same charge distribution on the plate as that given by the starting point, Eq. (9.1). We checked the calculations by a similar procedure in the other cases of two and four plates.

The equipotentials given by Eq. (9.13) are shown in Figure 9.2 in approximation (A) with $\ell_y/\ell_z = 3$, $\phi(0, 0, -\ell_z/2) = \phi_0/2$ and $\phi(0, 0, \ell_z/2) = -\phi_0/2$. In this case $\sigma_B = -2\pi\varepsilon_0\phi_0/\ell_z$ and $\sigma_A = 0$.

The lines of electric field are given by a function $\xi(x, 0, z)$ such that $\nabla\xi \cdot \nabla\phi = 0$. Following the procedure described in Section 6.5 we obtain in this case:

$$\xi(x, 0, z) = \frac{2\sigma_A z + \sigma_B(4\lambda_1 x - x^2 + z^2)/\ell_z}{\varepsilon_0}, \text{ if } x > 0, \qquad (9.15)$$

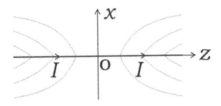

Figure 9.2: Equipotential lines in a plane orthogonal to the plate. The arrows indicate the direction of the current.

$$\xi(x, 0, z) = \frac{2\sigma_A z - \sigma_B(4\lambda_1 x + x^2 - z^2)/\ell_z}{\varepsilon_0}, \text{ if } x < 0. \quad (9.16)$$

This function presents a family of two hyperbolas in the regions above and below the plate. An example of this function ξ is presented in Figure 9.3 in approximation (A) with $\ell_y/\ell_z = 3$, $\phi(0,0,-\ell_z/2) = \phi_0/2$ and $\phi(0,0,\ell_z/2) = -\phi_0/2$.

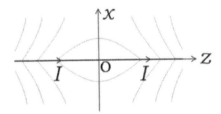

Figure 9.3: Lines of electric field in a plane orthogonal to the plate. The arrows indicate the direction of the current.

9.3 Two Parallel Plates

We now consider the experiments of Bergmann, Schaefer and Jefimenko utilizing a different model. We first consider a single straight conductor, Figures 3.2 and 3.3. Here we model these cases as that of a constant current flowing uniformly along the z axis of a conductor of conductivity g in the form of a parallelepiped of lengths ℓ_y, $2a$ and ℓ_z. Accordingly there will be free charges only along its outer surfaces located at $x = \pm a$ (considering the thick conductor centered at $(x, y, z) = (0, 0, 0)$). At both sides the free charges will be given by Equation (9.1). The superposition of the two charged planes situated in $x = a$ and $x = -a$, utilizing Eq. (9.13) and replacing x by $x \pm a$ appropriately yields the potential in the plane $y = 0$ as given by:

$$\phi(x, 0, z) = \frac{1}{\varepsilon_0}\left[\left(\sigma_A + \sigma_B \frac{z}{\ell_z}\right)\left(2\lambda_1 - \frac{|x-a|+|x+a|}{2}\right) + 2\sigma_A\lambda_2\right]. \quad (9.17)$$

This potential can be seen in Figure 9.4 in approximation (A) with $\ell_y/\ell_z = \ell_z/2a = 6.5$. With the boundary conditions $\phi(\pm a,\, 0,\, -\ell_z/2) = \phi_0/2$ and $\phi(\pm a,\, 0,\, \ell_z/2) = -\phi_0/2$ we have $\sigma_B = -\phi_0 \varepsilon_0/(4\lambda_1 - 2a)$ and $\sigma_A = 0$.

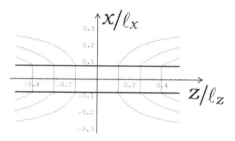

Figure 9.4: Equipotential lines in a plane orthogonal to two parallel plates carrying steady currents along the positive z direction.

The electric field is readily given by $\vec{E} = -\nabla \phi$:

$$\vec{E}(x > a,\, 0,\, z) = \frac{1}{\varepsilon_0}\left[\left(\sigma_A + \sigma_B \frac{z}{\ell_z}\right)\hat{x} - \sigma_B \frac{2\lambda_1 - x}{\ell_z}\hat{z}\right], \quad (9.18)$$

$$\vec{E}(-a < x < a,\, 0,\, z) = -\frac{1}{\varepsilon_0}\sigma_B \frac{2\lambda_1 - a}{\ell_z}\hat{z}, \quad (9.19)$$

$$\vec{E}(x < -a,\, 0,\, z) = -\frac{1}{\varepsilon_0}\left[\left(\sigma_A + \sigma_B \frac{z}{\ell_z}\right)\hat{x} - \sigma_B \frac{2\lambda_1 + x}{\ell_z}\hat{z}\right]. \quad (9.20)$$

As expected, the electric field is constant in the region between the two plates. This fact allows us to utilize the situation of two plates to model also the parallelepiped of sides ℓ_y and $2a$ carrying a steady current along the z direction. The two plates already mentioned would be equivalent to the top and bottom plates of the parallelepiped located in the planes $x = \pm a$.

The lines of electric field $\xi(x,\, 0,\, z)$ such that $\nabla \xi \cdot \nabla \phi = 0$ can be obtained by the method described before. They are given by the following equation:

$$\xi(x,\, 0,\, z) = \begin{cases} (2\sigma_A z + \sigma_B(4\lambda_1 x - x^2 + z^2)/\ell_z)/\varepsilon_0, & x > a, \\ -\sigma_B a x/\ell_z \varepsilon_0, & -a < x < a, \\ (2\sigma_A z - \sigma_B(4\lambda_1 x + x^2 - z^2)/\ell_z)/\varepsilon_0, & x < -a. \end{cases} \quad (9.21)$$

In Figure 9.5 we plot this function with the approximation $\ell_y/\ell_z = \ell_z/2a = 6.5$, in order to have similar dimensions as in Jefimenko's experiment. This theoretical Figure is similar to Jefimenko's experimental one, Figure 3.3.

9.4 Four Parallel Plates

We now wish to obtain plots similar to Figures 3.5 and 3.2 utilizing the parallelepiped model of this Chapter. We have essentially a transmission line in which

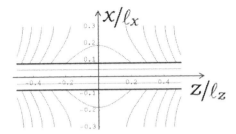

Figure 9.5: Lines of electric field in a plane orthogonal to two parallel plates carrying steady currents along the positive z direction.

the current flows uniformly along the z axis of a parallelepiped of conductivity g_1 and thickness $2a$, returning uniformly along another parallel parallelepiped of the same thickness but conductivity g_2. The centers of the two conductors are separated by a distance b. In this case there will be free charges in the four planes situated at $y = b/2 \pm a$ and $y = -b/2 \pm a$, with $b/2 > a > 0$.

9.4.1 Opposite Potentials

In this case both conductors have the same finite conductivity $g_1 = g_2 = g$. We assume that the potentials are exactly opposite in the two thick plates, for any z. The densities of surface charges for the plates located at $x = \pm(b/2+a)$ and $x = \pm(b/2-a)$ are given by:

$$\sigma(x = \pm(b/2+a),\, y,\, z) = \pm \left(\sigma_{A\text{ext}} + \sigma_{B\text{ext}} \frac{z}{\ell_z} \right), \tag{9.22}$$

$$\sigma(x = \pm(b/2-a),\, y,\, z) = \pm \left(\sigma_{A\text{int}} + \sigma_{B\text{int}} \frac{z}{\ell_z} \right). \tag{9.23}$$

We can obtain the potential utilizing Eq. (9.3). To simplify the results we define two dimensionless constants with appropriate values for each one of the approximations (Eq. (9.4) to (9.6)), namely:

$$(A) \quad \kappa_1 \equiv \frac{4b - 8a}{\pi \ell_z - 4b + 8a}, \quad \kappa_2 \equiv \frac{2b - 4a}{\pi \ell_z - 2b + 4a}, \tag{9.24}$$

$$(B) \quad \kappa_2 \equiv \frac{3\sqrt{2}(b - 2a)}{\pi \ell_z - 3\sqrt{2}(b - 2a)}, \quad \kappa_2 \equiv \frac{2\sqrt{2}(b - 2a)}{\pi \ell_z - 2\sqrt{2}(b - 2a)}, \tag{9.25}$$

$$(C) \quad \kappa_1 \equiv \frac{2b - a}{\pi \ell_z - 2b + a}, \quad \kappa_2 \equiv \frac{2b - a}{\pi(\pi \ell_z - 2b + a)}. \tag{9.26}$$

With the given approximations we have $\kappa_1 \ll 1$ and $\kappa_2 \ll 1$.

In order to model the given experiments, the potential should not depend on x in the regions $b/2 - a < x < b/2 + a$ and $-b/2 - a < x < -b/2 + a$ (as the current flows only along the z direction in these regions). This yields

$$\sigma_{Aint} = \frac{\sigma_{Aext}}{\kappa_2} \equiv \sigma_A \;,\quad \sigma_{Bint} = \frac{\sigma_{Bext}}{\kappa_1} \equiv \sigma_B \;. \tag{9.27}$$

The potential is then given by (in the plane $y = 0$ and in the following regions, respectively: $x > b/2 + a$, $b/2 - a < x < b/2 + a$, $-b/2 + a < x < b/2 - a$, $-b/2 - a < x < -b/2 + a$, $x < -b/2 - a$):

$$\phi = \begin{cases} ((b - 2a)(\sigma_A + \sigma_B z/\ell_z) + (b + 2a - y)(\sigma_A \kappa_2 + \sigma_B z \kappa_1/\ell_z))/2\varepsilon_0 \;, \\ (b - 2a)(\sigma_A + \sigma_B z/\ell_z)/2\varepsilon_0 \;, \\ y(\sigma_A + \sigma_B z/\ell_z)/\varepsilon_0 \;, \\ -(b - 2a)(\sigma_A + \sigma_B z/\ell_z)/2\varepsilon_0 \;, \\ -((b - 2a)(\sigma_A + \sigma_B z/\ell_z) + (b + 2a + y)(\sigma_A \kappa_2 + \sigma_B z \kappa_1/\ell_z))/2\varepsilon_0 \;. \end{cases} \tag{9.28}$$

This potential can be seen in Figure 9.6 in approximation (A) with $\ell_y/\ell_z = \ell_z/2a = 6.8$.

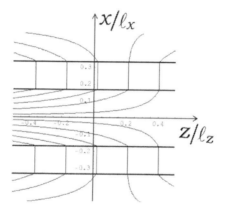

Figure 9.6: Equipotential lines in a plane orthogonal to the four plates. There is a current along the positive (negative) z direction in the two top (bottom) plates.

The electric field $\vec{E}(x, 0, z) = -\nabla \phi$ is given in the five regions by, respectively:

$$\vec{E} = \begin{cases} ((\sigma_A \kappa_2 + \sigma_B z \kappa_1/\ell_z)\hat{x} - \sigma_B[b - 2a + \kappa_1(b + 2a - 2x)]\hat{z}/\ell_z)/2\varepsilon_0 \;, \\ -(b - 2a)\sigma_B \hat{z}/2\varepsilon_0 \ell_z \;, \\ -((\sigma_A + \sigma_B z/\ell_z)\hat{x} + x\sigma_B \hat{z}/\ell_z)/\varepsilon_0 \;, \\ (b - 2a)\sigma_B \hat{z}/2\varepsilon_0 \ell_z \;, \\ ((\sigma_A \kappa_2 + \sigma_B z \kappa_1/\ell_z)\hat{x} + \sigma_B[b - 2a + \kappa_1(b + 2a + 2x)]\hat{z}/\ell_z)/2\varepsilon_0 \;. \end{cases} \tag{9.29}$$

The lines of electric field, $\xi(x, 0, z)$, are given for each region in Eq. (9.30):

$$\xi = \begin{cases} (2\sigma_A z \kappa_2/\kappa_1 + \sigma_B[[b + 2a + (b - 2a)/\kappa_1]x - x^2 + z^2]/\ell_z)/\varepsilon_0, \\ -\sigma_B(b - 2a)x/2\ell_z\varepsilon_0, \\ (2\sigma_A z - \sigma_B(x^2 - z^2)/\ell_z)/\varepsilon_0, \\ \sigma_B(b - 2a)x/2\ell_z\varepsilon_0, \\ (2\sigma_A z \kappa_2/\kappa_1 - \sigma_B[[b + 2a + (b - 2a)/\kappa_1]x - x^2 + z^2]/\ell_z)/\varepsilon_0. \end{cases} \quad (9.30)$$

In Figure 9.7 we plot this function in the approximation (A) with $\ell_y/\ell_z = \ell_z/2a = 6.8$. The upper plate has the potential at its boundaries given by $\phi(b/2 - a < x < b/2 + a, 0, -\ell_z/2) = \phi_0/2$ and $\phi(b/2 - a < x < b/2 + a, 0, \ell_z/2) = 0$, while the lower plate has the potential at its boundaries given by $\phi(-b/2 - a < x < -b/2 + a, 0, -\ell_z/2) = -\phi_0/2$ and $\phi(-b/2 - a < x < -b/2 + a, 0, \ell_z/2) = 0$.

The relation between ϕ_0 and the surface charges for this case is given by $\sigma_A = \varepsilon_0\phi_0/2(b - 2a)$ and $\sigma_B = -\varepsilon_0\phi_0/(b - 2a)$.

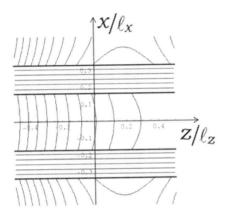

Figure 9.7: Lines of electric field in a plane orthogonal to the four plates. There is a current along the positive (negative) z direction in the two top (bottom) plates.

9.4.2 Perfect Conductor Plate

Now, suppose that the two lower plates (or the lower parallelepiped) are a perfect conductor, with zero resistivity. That is, suppose they are subjected to the same constant potential $\phi(-b/2 - a < x < -b/2 + a, 0, z) = \Phi$ in the whole extension along the z axis, but still conducting a steady current. This experimental result is shown at the right side of Figure 3.5 with $g_1 \ll g_2$. To model this case we consider four plates located at $x = b/2 + a$, $x = b/2 - a$, $x = -b/2 + a$ and $x = -b/2 - a$. Their surface charges are given by, respectively, $\sigma(x = b/2 + a, y, z) = \sigma_{Ab} + \sigma_{Bb}z/\ell_z$, $\sigma(x = b/2 - a, y, z) = \sigma_{Aa} + \sigma_{Ba}z/\ell_z$,

$\sigma(x = -b/2 + a, y, z) = \sigma_{-Aa} + \sigma_{-Ba} z/\ell_z$ and $\sigma(x = -b/2 - a, y, z) = \sigma_{-Ab} + \sigma_{-Bb} z/\ell_z$.

The potential must not depend on x in the region $b/2 - a < x < b/2 + a$, and must be a constant in the region $-b/2 - a < x < -b/2 + a$. From this we find:

$$\sigma_{Aa} = \frac{\sigma_{Ab}(4\lambda_1 + 4\lambda_2 - b - 2a) - 2\Phi\varepsilon_0}{b - 2a}, \qquad \sigma_{Ba} = \sigma_{Bb}\frac{4\lambda_1 - b - 2a}{b - 2a},$$

$$\sigma_{-Aa} = -\sigma_{Aa}, \qquad \sigma_{-Ba} = -\sigma_{Ba},$$

$$\sigma_{-Ab} = \sigma_{Ab}, \qquad \sigma_{-Bb} = \sigma_{Bb}. \tag{9.31}$$

With Eq. (9.13) and the appropriate replacements of x by $x \pm (b/2 \pm a)$ we get in the five regions, respectively:

$$\phi(x, 0, z) = \begin{cases} [(\sigma_{Ab} + \sigma_{Bb} z/\ell_z)(4\lambda_1 - b/2 - a - y) + 4\lambda_2 \sigma_{Ab}]/\varepsilon_0 - \Phi, \\ (2b - 4a)(\sigma_{Aa} + \sigma_{Ba} z/\ell_z)/\varepsilon_0 + \Phi, \\ (\sigma_{Aa} + \sigma_{Ba} z/\ell_z)(b/2 - a + y)/\varepsilon_0 + \Phi, \\ \Phi, \\ (\sigma_{Ab} + \sigma_{Bb} z/\ell_z)(b/2 + a + y)/\varepsilon_0 + \Phi. \end{cases}$$
$$\tag{9.32}$$

The equipotentials are shown in Figure 9.8 in approximation (A) with $\ell_y/\ell_z = \ell_z/2a = 6.8$.

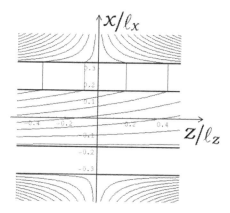

Figure 9.8: Equipotential lines in a plane orthogonal to the four plates. The two top plates are uniformly resistive and carry currents along the positive z direction. The two bottom plates have zero resistivity and carry currents along the negative z direction.

The electric field in these five regions is given by, respectively:

$$\vec{E} = \begin{cases} [(\sigma_{Ab} + \sigma_{Bb}z/\ell_z)\hat{x} - \sigma_{Bb}(4\lambda_1 - b/2 - a - x)\hat{z}/\ell_z]/\varepsilon_0 \,, \\ -(b-2a)\sigma_{Ba}\hat{z}/\ell_z\varepsilon_0 \,, \\ -[(\sigma_{Aa} + \sigma_{Ba}z/\ell_z)\hat{x} + \sigma_{Ba}(b/2 - a + x)\hat{z}/\ell_z]/\varepsilon_0 \,, \\ \vec{0} \,, \\ -[(\sigma_{Ab} + \sigma_{Bb}z/\ell_z)\hat{x} + \sigma_{Bb}(b/2 + a + x)\hat{z}/\ell_z]/\varepsilon_0 \,. \end{cases} \quad (9.33)$$

The lines of electric field are given by:

$$\xi(x,\,0,\,z) = \begin{cases} (2\sigma_{Ab}z + \sigma_{Bb}((8\lambda_1 - b - 2a)x - x^2 + z^2)/\ell_z)/\varepsilon_0 \,, \\ -\sigma_{Bb}(b-2a)x/2\ell_z\varepsilon_0 \,, \\ (2\sigma_{Aa}z - \sigma_{Ba}((b - 2a)x + x^2 - z^2)/\ell_z)/\varepsilon_0 \,, \\ -\sigma_{Ba}(b-2a)^2/4\ell_z\varepsilon_0 \,, \\ (2\sigma_{Ab}z - \sigma_{Bb}((b + 2a)x + x^2 - z^2)/\ell_z)/\varepsilon_0 \,. \end{cases} \quad (9.34)$$

They are shown in Figure 9.9 with the given approximation and the same dimensions as in Figure 9.7. The constant potential in the lower plate is $\Phi = -\phi_0/2$. Once more there is a reasonable match with Jefimenko's experimental result, the right side of Figure 3.5, especially in the region between the parallelepipeds.

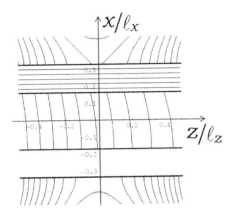

Figure 9.9: Lines of electric field in a plane orthogonal to the four plates. The two top plates are uniformly resistive and carry currents along the positive z direction. The two bottom plates have zero resistivity and carry currents along the negative z direction.

Chapter 10

Resistive Strip

10.1 The Problem

Here we consider a constant current flowing uniformly through the surface of a stationary and resistive straight strip. Our goal is to calculate the potential ϕ and electric field $\vec{E}$ everywhere in space and the surface charge distribution σ along the strip that creates this electric field. We follow essentially the work published in 2003 [223].

We consider a strip in the $x = 0$ plane localized in the region $-a < y < a$ and $-\ell/2 < z < \ell/2$, such that $\ell \gg a > 0$. The medium around the strip is taken to be air or vacuum. The constant current I flows uniformly along the positive z direction with a surface current density given by $\vec{K} = I\hat{z}/2a$ (see Fig. 10.1). By Ohm's law this uniform current distribution is related to a spatially constant electric field along the surface of the strip. In the steady state this electric field can be related to the potential by $\vec{E} = -\nabla \phi$. This relation means that along the strip the potential is a linear function of z and independent of y. The problem can then be solved by finding the solution of Laplace's equation $\nabla^2 \phi = 0$ in empty space and applying the boundary conditions.

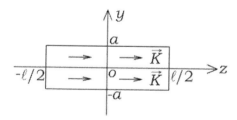

Figure 10.1: A resistive strip of width $2a$ and length ℓ with a steady and uniform surface current density $\vec{K}$ along the positive z direction.

10.2 The Solution

Due to the symmetry of the problem, it is convenient to utilize elliptic-cylindrical coordinates (ζ, ϑ, z) see Figure 10.2 [224]. These variables can take the following values: $0 \leq \zeta \leq \infty$, $0 \leq \vartheta \leq 2\pi$ rad, and $-\infty \leq z \leq \infty$. The relation between cartesian (x, y, z) and elliptic-cylindrical coordinates is given by:

$$x = a \sinh \zeta \sin \vartheta , \qquad (10.1)$$

$$y = a \cosh \zeta \cos \vartheta , \qquad (10.2)$$

$$z = z , \qquad (10.3)$$

where $2a$ is the constant thickness of the strip. The inverse relations are given by:

$$\zeta = \tanh^{-1} \sqrt{\frac{y^2 - x^2 - a^2 + \Omega}{2y^2}} , \qquad (10.4)$$

$$\vartheta = \tan^{-1} \sqrt{\frac{a^2 + x^2 - y^2 + \Omega}{2y^2}} , \qquad (10.5)$$

$$z = z , \qquad (10.6)$$

where $\Omega \equiv \sqrt{(x^2 + y^2 + a^2)^2 - 4a^2 y^2}$.

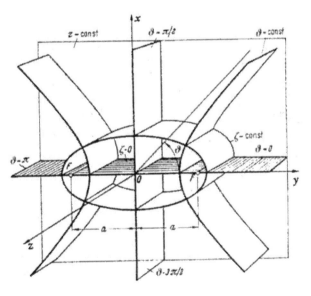

Figure 10.2: Elliptic-cylindrical coordinates (ζ, ϑ, z).

Laplace's equation in this coordinate system is given by:

$$\nabla^2 \phi = \frac{1}{a^2(\cosh^2 \zeta - \cos^2 \vartheta)} \left(\frac{\partial^2 \phi}{\partial \zeta^2} + \frac{\partial^2 \phi}{\partial \vartheta^2} \right) + \frac{\partial^2 \phi}{\partial z^2} = 0 . \quad (10.7)$$

A solution of Eq. (10.7) can be obtained by separation of variables in the form $\phi(\zeta, \vartheta, z) = H(\zeta)\Phi(\vartheta)Z(z)$:

$$H'' - (\alpha_2 + \alpha_3 a^2 \cosh^2 \zeta)H = 0 , \quad (10.8)$$

$$\Phi'' + (\alpha_2 + \alpha_3 a^2 \cos^2 \vartheta)\Phi = 0 , \quad (10.9)$$

$$Z'' + \alpha_3 Z = 0 , \quad (10.10)$$

where α_2 and α_3 are constants.

For the long strip being considered here, it is possible to neglect boundary effects near $z = \pm \ell/2$. It has already been proved that in this case the potential must be a linear function of z, not only over the strip, but also over all space. See Section 5.4. This condition means that $\alpha_3 = 0$. There are then two possible solutions for $\Phi(\vartheta)$. If $\alpha_2 = 0$, then $\Phi = C_1 + C_2 \vartheta$; if $\alpha_2 \neq 0$, then $\Phi = C_3 \sin(\sqrt{\alpha_2}\vartheta) + C_4 \cos(\sqrt{\alpha_2}\vartheta)$, where C_1 to C_4 are constants. Along the strip we have $x = 0$, and $y^2 \leq a^2$, which means that $\Omega = a^2 - y^2$, $\zeta = 0$ and $\vartheta = \tan^{-1}\sqrt{(a^2 - y^2)/y^2}$. We are assuming that the potential does not depend on y along the strip. This independence and the relation between y and ϑ means that the potential will not depend on ϑ as well. Thus a non-trivial solution for Φ can only exist if $\alpha_2 = 0$, $C_2 = 0$, and $\Phi = $ constant for all ϑ. The solution for H with $\alpha_2 = \alpha_3 = 0$ will be then a linear function of ζ. The general solution of the problem is then given by:

$$\phi = (A_1 \zeta - A_2)\left(\phi_A + \phi_B \frac{z}{\ell}\right)$$

$$= \left(A_1 \tanh^{-1}\sqrt{\frac{y^2 - x^2 - a^2 + \Omega}{2y^2}} - A_2 \right)\left(\phi_A + \phi_B \frac{z}{\ell}\right) . \quad (10.11)$$

The electric field $\vec{E} = -\nabla \phi$ takes the form:

$$\begin{aligned}\vec{E} =\ & -A_1 \left(\frac{|y|x\sqrt{2}}{\Omega\sqrt{y^2 - x^2 - a^2 + \Omega}} \hat{x} \right. \\ & \left. + \frac{|y|\sqrt{y^2 - x^2 - a^2 + \Omega}}{y\sqrt{2}\Omega} \hat{y} \right)\left(\phi_A + \phi_B \frac{z}{\ell}\right) \\ & - \frac{\phi_B}{\ell}\left(A_1 \tanh^{-1}\sqrt{\frac{y^2 - x^2 - a^2 + \Omega}{2y^2}} - A_2 \right)\hat{z} , \quad (10.12)\end{aligned}$$

To find the surface charge density, we utilize the approximation close to the strip ($|y| < a$ and $|x| \ll a$):

$$\vec{E} \approx -A_1 \left[\frac{x}{|x|\sqrt{a^2-y^2}} \hat{x} + \frac{y|x|}{(a^2-y^2)^{3/2}} \hat{y} \right] \left(\phi_A + \phi_B \frac{z}{\ell} \right)$$

$$- \frac{\phi_B}{\ell} \left(A_1 \tanh^{-1} \frac{|x|}{\sqrt{a^2-y^2}} - A_2 \right) \hat{z} . \quad (10.13)$$

The surface charge density $\sigma(y, z)$ can be obtained by the standard procedure utilizing Gauss's law $\oiint_S \vec{E} \cdot d\vec{a} = Q/\varepsilon_0$. The surface charge density is then obtained by considering the limit in which $|x| \to 0$ in Eq. (10.13) and a small cylindrical volume with its length much smaller than its diameter, yielding: $\sigma = \varepsilon_0[\vec{E}(x>0) \cdot \hat{x} - \vec{E}(x<0) \cdot (-\hat{x})]$. If we use Eq. (10.13), the surface charge density is found to be given by:

$$\sigma(x, z) = -\frac{2\varepsilon_0 A_1(\phi_A + \phi_B z/\ell)}{\sqrt{a^2-y^2}} . \quad (10.14)$$

The linear charge density $\lambda(z)$ can be obtained as $\lambda(z) = \int_{-a}^{a} \sigma(y, z) dy$, yielding

$$\lambda(z) = -2\pi\varepsilon_0 A_1 \left(\phi_A + \phi_B \frac{z}{\ell} \right) . \quad (10.15)$$

10.3 Discussion

In the plane $x = 0$ the current in the strip creates a magnetic field $\vec{B}$ that points along the positive (negative) x direction for $y > 0$ ($y < 0$). Consider a specific conduction electron moving with drifting velocity $\vec{v}_d$. The magnetic field due to all other mobile conduction electrons will act on this specific conduction electron with a force given by $q\vec{v}_d \times \vec{B}$ (see Fig. 10.3). This force will cause a redistribution of charges along the y direction, with negative charges concentrating along the center of the strip and positive charges at the extremities $y = \pm a$. In the steady-state this redistribution of charges will create an electric field along the y direction, E_y, that will balance the magnetic force, namely, $|qE_y| = |qv_d B|$.

We have disregarded this Hall electric field because it is usually much smaller than the electric field giving rise to the current, as was shown in 6.4.

We now analyze some particular cases. We first consider two limits by comparing a with the distance of the observation point $\rho = \sqrt{x^2 + y^2}$. If $a^2 \gg \rho^2$, we have $\Omega \approx a^2 + x^2 - y^2 + 2x^2y^2/a^2$ and $\zeta \approx |x|/a$, such that:

$$\phi \approx \left(A_1 \frac{|x|}{a} - A_2 \right) \left(\phi_A + \phi_B \frac{z}{\ell} \right) . \quad (10.16)$$

Combining this result with Eq. (10.14) in the approximation $a^2 \gg \rho^2$ yields:

> Z

Figure 10.3: Magnetic field $\vec{B}$ along the strip due to the current along the positive z direction. There is a magnetic force pointing toward the axis acting upon the conduction electrons.

$$\phi \approx \frac{\sigma(z)}{2\varepsilon_0} \left(a \frac{A_2}{A_1} - |x| \right) . \tag{10.17}$$

This result coincides with Eq. (9.9) considering $aA_2/A_1 = (\ell_y/\pi) \ln(2\ell_z/\ell_y)$ and $\sigma_A = 0$. This was expected because a strip with $\ell^2 \gg a^2 \gg \rho^2$ is equivalent to a large plate.

This Equation is also equivalent to Eq. (6.26) with $A_2/A_1 = \ln(\ell/a)$ except by an overall factor of 2. This was once more to be expected. The electrostatic potential at a distance d from a charged plate is given by $\phi = \phi_0 - \sigma d/2\varepsilon_0$, where ϕ_0 is an arbitrary constant and σ is the total density of surface charge, half of it in each side of the charged plane. On the other hand, the electrostatic potential just outside a closed charged conductor (at a distance d from it) is given $\phi = \phi_1 - \sigma d/\varepsilon_0$, where ϕ_1 is an arbitrary constant and σ here is the surface charge density at the point in which the potential is being estimated, while the internal potential has the constant value ϕ_1. That is, when we close an open charged conducting surface, the charges in the internal side migrate to the external side. The magnitude of the electric field just outside the closed surface is twice the electric field close to a large charged plane, supposing the same local charge density in both cases. As we have seen here, the same happens with the surface charges when a current flows along the resistive surface.

On the other hand, if $a^2 \ll \rho^2$, we have $\Omega \approx \rho^2 + a^2 - 2a^2y^2/\rho^2$ and $\zeta \approx \ln(\rho/a)$. Utilizing these results in Eq. (10.11) combined with Eq. (10.15) yields:

$$\phi \approx \frac{\lambda(z)}{2\pi\varepsilon_0} \left(\frac{A_2}{A_1} - \ln \frac{\rho}{a} \right) . \tag{10.18}$$

This result coincides with Eq. (6.11) with $A_2/A_1 = \ln(\ell/a)$, where ℓ is the typical length of the wire or strip being considered, with $\ell \gg a$. This is reasonable because Eq. (6.11) corresponds to the potential outside a long straight cylindrical wire carrying a constant current. At a point far from the axis of the strip both results coincide as they must.

10.4 Comparison with the Experimental Results

These results indicate that there is an electric field not only along the resistive strip carrying a steady current, but also in the space around the strip. As we have seen, Jefimenko performed experiments which demonstrate the existence of this external electric field [174] [176, Plate 6]. The configuration of Jefimenko's experiment, Figure 3.3, is equivalent to what has been considered here: a two-dimensional conducting strip made on a glass plate using a transparent conducting ink. To compare our calculations with his experimental results, we need the values of A_2/A_1 and ϕ_A/ϕ_B. We take $A_2/A_1 = 3.6$ and $\phi_A/\phi_B = 0$. The condition $\phi_A/\phi_B = 0$ corresponds to the symmetrical case considered by Jefimenko in which the electric field is parallel to the conductor just outside of it at $z = 0$ (zero density of surface charges at $z = 0$).

We first consider the plane orthogonal to the strip, $y = 0$. In this case the potential reduces to:

$$\phi(x, 0, z) = \left(A_1 \tanh^{-1} \sqrt{\frac{x^2}{x^2 + a^2}} - A_2 \right) \left(\phi_A + \phi_B \frac{z}{\ell} \right) . \tag{10.19}$$

The lines of the electric field orthogonal to the equipotentials can be obtained by the procedure described in Section 6.5. These lines are represented by a function ξ such that $\nabla \xi \cdot \nabla \phi = 0$. This yields:

$$\begin{aligned}
\xi(x, 0, z) = {} & A_1 \phi_B \frac{z^2}{\ell^2} + 2 A_1 \phi_A \frac{z}{\ell} + \frac{A_1 \phi_B}{2} \frac{x^2}{\ell^2} \\
& - A_1 \phi_B \frac{|x|}{\ell} \frac{\sqrt{x^2 + a^2}}{\ell} \cosh^{-1} \sqrt{\frac{x^2 + a^2}{a^2}} \\
& - \frac{A_1 \phi_B}{2} \frac{a^2}{\ell^2} \left(\cosh^{-1} \sqrt{\frac{x^2 + a^2}{a^2}} \right)^2 \\
& - \frac{A_2 \phi_B}{4} \left(\frac{|x|}{\ell} \frac{\sqrt{x^2 + a^2}}{\ell} + \frac{a^2}{\ell^2} \ln \frac{|x| + \sqrt{x^2 + a^2}}{a} \right) . \tag{10.20}
\end{aligned}$$

A plot of Eqs. (10.19) and (10.20) is given in Fig. 10.4.

We now consider the plane of the strip, $x = 0$. The potential reduces to:

$$\phi(0, |y| \leq a, z) = -A_2 \left(\phi_A + \phi_B \frac{z}{\ell} \right) , \tag{10.21}$$

$$\begin{aligned}
\phi(0, |y| \geq a, z) &= \left(A_1 \tanh^{-1} \sqrt{\frac{y^2 - a^2}{y^2}} - A_2 \right) \left(\phi_A + \phi_B \frac{z}{\ell} \right) \\
&= \left(A_1 \cosh^{-1} \frac{|y|}{a} - A_2 \right) \left(\phi_A + \phi_B \frac{z}{\ell} \right) . \tag{10.22}
\end{aligned}$$

When there is no current in the strip, the potential along it is a constant for all z. From Eq. (10.21) this means $\phi_B = 0$. This value of ϕ_B in Eqs. (10.11),

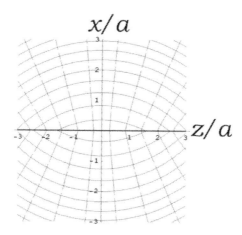

Figure 10.4: Equipotentials (dashed lines) and lines of electric field (continuous lines) in a plane orthogonal to the plane of the strip.

(10.12) and (10.14) reduces these equations to the known electrostatic solution of a strip charged to a constant potential [225].

By a similar procedure, the lines of electric field for the plane $x = 0$ are given by:

$$\xi(0, |y| \leq a, z) = A_2 \phi_B \frac{ay}{\ell^2}, \tag{10.23}$$

$$\xi(0, |y| \geq a, z) = A_1 \phi_B \frac{z^2}{\ell^2} + 2 A_1 \phi_A \frac{z}{\ell} + \frac{A_1 \phi_B}{2} \frac{y^2}{\ell^2}$$

$$- A_1 \phi_B \frac{|y|}{\ell} \frac{\sqrt{y^2 - a^2}}{\ell} \cosh^{-1} \frac{|y|}{a}$$

$$+ \frac{A_1 \phi_B}{2} \frac{a^2}{\ell^2} \left(\cosh^{-1} \frac{|y|}{a} \right) - \frac{A_2 \phi_B}{4} \left(\frac{|y|}{\ell} \frac{\sqrt{y^2 - a^2}}{\ell} \right.$$

$$\left. - \frac{a^2}{\ell^2} \ln \frac{|y| + \sqrt{y^2 - a^2}}{a} \right). \tag{10.24}$$

A plot of Eqs. (10.21) to (10.24) is presented in Fig. 10.5.

Figure 10.6 presents the theoretical electric field lines overlaid on the experimental result of Jefimenko, Figure 3.3.

In Fig. 10.7 the experimental result of Jefimenko, Barnett and Kelly, Figure 3.10, is overlaid on the equipotential lines calculated utilizing Eqs. (10.23) and (10.24) with $A_2/A_1 = 3.0$ and $\phi_A/\phi_B = 0$. The agreement is not as good

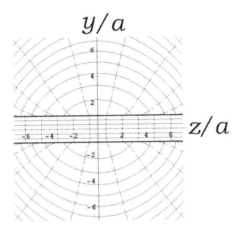

Figure 10.5: Equipotentials (dashed lines) and lines of electric field (continuous lines) in the plane of the strip.

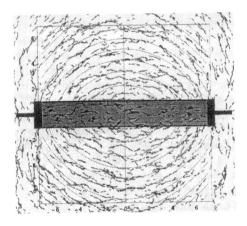

Figure 10.6: Theoretical lines of electric field overlaid on the experimental lines obtained by Jefimenko.

as in our previous figure for two reasons: One reason is that our calculations are for a two-dimensional configuration, while the experiment of Jefimenko, Barnett and Kelly [177] was performed in a three-dimensional rectangular chamber. The second reason is that in the grass seed experiment [174], the ratio of the length to the width of the conductor was 7, but in the second experiment [177], this ratio was only 2, which means that boundary effects near $z = \ell/2$ and $z = -\ell/2$ are more important. These boundary effects were not considered in our calculations.

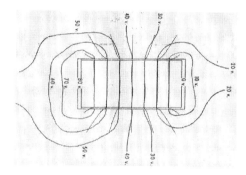

Figure 10.7: Theoretical equipotential lines overlaid on the experimental lines obtained by Jefimenko, Barnett and Kelly.

Part III

Curved Conductors

In this third Part of the book we consider resistive conductors carrying steady currents along curved paths. Russell's theorem, discussed in Section 5.4, is no longer valid due to the curvature of the wire. Three cases in particular will be discussed here, the azimuthal current in an infinite cylindrical shell, the azimuthal current over the surface of a spherical shell, and the azimuthal current in a toroidal conductor. These cases can still be solved analytically, and their solutions clarify some important aspects of surface charges in conductors carrying steady currents.

Chapter 11

Resistive Cylindrical Shell with Azimuthal Current

11.1 Configuration of the Problem

The subject of this Chapter has been discussed mainly by Jefimenko [175, Problem 9.33 and Figure 14.7] [176, p. 318], Heald [226] and Griffiths [16, p. 279]. We follow these approaches here.

An infinite homogeneous resistive cylindrical shell of radius a has its axis coinciding with the z direction. We utilize cylindrical coordinates (ρ, φ, z) with origin at the center of the shell, with $\rho = \sqrt{x^2 + y^2}$ being the distance to the z axis. There is a narrow slot along its entire length located at $(\rho, \varphi, z) = (a, \pi, z)$. An idealized line battery in the slot maintains its terminals located at $\varphi = \pm \pi$ rad with the constant potentials $\phi = \pm \phi_B/2$, respectively. See Figure 11.1.

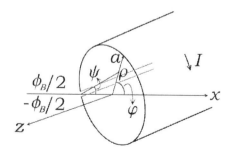

Figure 11.1: Configuration of the problem.

In accordance with Ohm's law the potential along the surface of the cylindrical shell is then given by

$$\phi(a, \varphi, z) = \phi_B \frac{\varphi}{2\pi} . \qquad (11.1)$$

11.2 Potential and Electric Field

The potential inside and outside the cylindrical shell satisfies Laplace's equation $\nabla^2 \phi = 0$:

$$\frac{1}{\rho}\frac{\partial}{\partial \rho}\left(\rho \frac{\partial \phi}{\partial \rho}\right) + \frac{1}{\rho^2}\frac{\partial^2 \phi}{\partial \varphi^2} + \frac{\partial^2 \phi}{\partial z^2} = 0 . \qquad (11.2)$$

The solution should be independent of z. Trying a solution in terms of separation of variables $\phi(\rho, \varphi, z) = R(\rho)\Phi(\varphi)$ yields

$$\Phi'' + m^2 \Phi = 0 , \qquad (11.3)$$

$$\rho \frac{d}{d\rho}\left(\rho \frac{dR}{d\rho}\right) - m^2 R = 0 , \qquad (11.4)$$

where m is a constant. The solutions of these equations if $m = 0$ are $\Phi(\varphi) = A_0 + B_0\varphi$ and $R(\rho) = C_0 \ln \rho + D_0$. If $m \neq 0$ we have $\Phi(\varphi) = A_m \cos(m\varphi) + B_m \sin(m\varphi)$ and $R(\rho) = C_m \rho^{-m} + D_m \rho^m$. The solutions must be periodic in φ, i.e., $\Phi(\varphi + 2\pi) = \Phi(\varphi)$. This means that $B_0 = 0$ and $m = 1, 2, 3, ...$

The internal and external solutions, with appropriate coefficients, are then given by

$$\phi(\rho \leq a, \varphi, z) = A_{0i}\left(C_{0i} \ln \rho + D_{0i}\right)$$
$$+ \sum_{m=1}^{\infty} \left[A_{mi} \cos(m\varphi) + B_{mi} \sin(m\varphi)\right] \left[C_{mi} \rho^{-m} + D_{mi} \rho^m\right] , \qquad (11.5)$$

$$\phi(\rho \geq a, \varphi, z) = A_{0e}\left(C_{0e} \ln \rho + D_{0e}\right)$$
$$+ \sum_{m=1}^{\infty} \left[A_{me} \cos(m\varphi) + B_{me} \sin(m\varphi)\right] \left[C_{me} \rho^{-m} + D_{me} \rho^m\right] , \qquad (11.6)$$

We specify finite solutions when $\rho \to 0$. This means that $C_{0i} = C_{1i} = C_{2i} = ... = 0$. We also specify solutions in which the potential goes to zero when $\rho \to \infty$. This means that $C_{0e} = D_{0e} = D_{1e} = D_{2e} = ... = 0$. Our solutions in these two regions reduce to

$$\phi(\rho \leq a, \varphi, z) = A_{0i}D_{0i} + \sum_{m=1}^{\infty} \left[A_{mi}D_{mi} \cos(m\varphi) + B_{mi}D_{mi} \sin(m\varphi)\right] \rho^m , \qquad (11.7)$$

$$\phi(\rho \geq a, \varphi, z) = \sum_{m=1}^{\infty} \left[A_{me} C_{me} \cos(m\varphi) + B_{me} C_{me} \sin(m\varphi) \right] \frac{1}{\rho^m} . \quad (11.8)$$

The potential must be continuous in $\rho = a$. This means that $A_{0i} D_{0i} = 0$, $A_{mi} D_{mi} a^m = A_{me} C_{me} a^{-m}$ and $B_{mi} D_{mi} a^m = B_{me} C_{me} a^{-m}$. Defining $A_{mi} D_{mi} \equiv G_m$ and $B_{mi} D_{mi} \equiv H_m$ yields

$$\phi(\rho \leq a, \varphi, z) = \sum_{m=1}^{\infty} \left[G_m \cos(m\varphi) + H_m \sin(m\varphi) \right] \rho^m , \quad (11.9)$$

$$\phi(\rho \geq a, \varphi, z) = \sum_{m=1}^{\infty} \left[G_m \cos(m\varphi) + H_m \sin(m\varphi) \right] \frac{a^{2m}}{\rho^m} . \quad (11.10)$$

Now we need to apply the boundary condition at $\rho = a$, Eq. (11.1). To this end we employ the Fourier expansion of φ, which is valid for $-\pi$ rad $< \varphi < \pi$ rad:

$$\varphi = 2 \left[\sum_{m=1}^{\infty} \frac{(-1)^{m-1} \sin(m\varphi)}{m} \right] . \quad (11.11)$$

Comparing Eqs. (11.9) and (11.10) at $\rho = a$ with Eqs. (11.1) and (11.11) yields $G_m = 0$ and $H_m a^m = \phi_B (-1)^{m+1}/\pi m$:

$$\phi(\rho \leq a, \varphi, z) = -\frac{\phi_B}{\pi} \left[\sum_{m=1}^{\infty} (-1)^m \left(\frac{\rho}{a} \right)^m \frac{\sin(m\varphi)}{m} \right] , \quad (11.12)$$

$$\phi(\rho \geq a, \varphi, z) = -\frac{\phi_B}{\pi} \left[\sum_{m=1}^{\infty} (-1)^m \left(\frac{a}{\rho} \right)^m \frac{\sin(m\varphi)}{m} \right] . \quad (11.13)$$

These two series can be put in closed form [226]:

$$\phi(\rho \leq a, \varphi, z) = \frac{\phi_B}{\pi} \tan^{-1} \frac{\rho \sin \varphi}{a + \rho \cos \varphi} = \frac{\phi_B}{\pi} \tan^{-1} \frac{y}{a + x} = \frac{\phi_B}{\pi} \psi , \quad (11.14)$$

$$\phi(\rho \geq a, \varphi, z) = \frac{\phi_B}{\pi} \tan^{-1} \frac{a \sin \varphi}{\rho + a \cos \varphi} = \frac{\phi_B}{\pi} \tan^{-1} \frac{ay}{x^2 + ax + y^2} , \quad (11.15)$$

where ψ is the polar angle about the slot as the axis.

The equipotentials given by these equations are represented in Figure 11.2.

The lines of electric field given by the function $\xi(x, y)$ such that $\nabla \xi \cdot \nabla \phi = 0$ can be obtained by the method described in Section 6.5. For the region $\rho < a$ this function is given by

$$\xi(x, y) = \phi_B \frac{2ax + x^2 + y^2}{a^2} . \quad (11.16)$$

These are circular arcs centered on the battery given by the following equation (for a particular ξ_o):

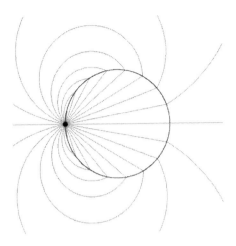

Figure 11.2: Equipotential lines. The battery is represented by the black spot.

$$(x+a)^2 + y^2 = \frac{2\xi_o + \phi_B}{\phi_B}a^2 \ . \tag{11.17}$$

Combining this result with Eqs. (8) and (10) of Heald's paper [226] we can also obtain the function ξ for the region outside the cylinder (this result can be checked by observing that it satisfies $\nabla \xi \cdot \nabla \phi = 0$):

$$\xi(x,y) = \phi_B \frac{a^2 + 2ax}{x^2 + y^2} \ . \tag{11.18}$$

These are also circular arcs with centers along the x axis, given by (for a particular ξ_o):

$$\left(x - \frac{\phi_B}{\xi_o}a\right)^2 + y^2 = \frac{\phi_B}{\xi_o}\frac{\xi_o + \phi_B}{\xi_o}a^2 \ . \tag{11.19}$$

From Eqs. (11.14), (11.15), (11.16) and (11.18) we can verify that $\nabla \xi \cdot \nabla \phi = 0$.

The electric field can be readily obtained by $\vec{E} = -\nabla \phi$. Inside the shell it is given by:

$$\vec{E}(\rho < a, \ \varphi, \ z) = -\frac{\phi_B}{\pi} \frac{a(\sin\varphi)\hat{\rho} + (\rho + a\cos\varphi)\hat{\varphi}}{a^2 + \rho^2 + 2a\rho\cos\varphi}$$

$$= -\frac{\phi_B}{\pi}\frac{\hat{\psi}}{\rho'} \ , \tag{11.20}$$

where $\rho' \equiv \sqrt{\rho^2 + a^2 + 2a\rho\cos\varphi}$ is the polar radius about the slot as the axis. That is, the lines of electric field are circular arcs centered on the battery.

Outside the shell the electric field is given by:

$$\vec{E}(\rho > a,\, \varphi,\, z) = \frac{\phi_B}{\pi} \frac{a}{\rho} \frac{\rho(\sin\varphi)\hat{\rho} - (a + \rho\cos\varphi)\hat{\varphi}}{a^2 + \rho^2 + 2a\rho\cos\varphi} \,, \tag{11.21}$$

with magnitude $|\vec{E}| = \phi_B a/\pi\rho\rho'$.

At the surface of the shell, $\rho = a$, Eqs. (11.20) and (11.21) yield the same tangential component:

$$E_\varphi(a, \varphi, z) = -\frac{\phi_B}{2\pi a} \,. \tag{11.22}$$

This is the correct result arising from Eq. (11.1), namely, $E = \Delta\phi/L$, where $\Delta\phi \equiv \phi_B$ is the potential difference generated by the battery and $L \equiv 2\pi a$ is the length described by the electrons around the circuit.

The lines of electric field are represented in Figure 11.3.

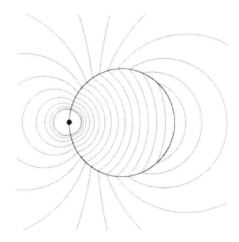

Figure 11.3: Lines of electric field. The battery is represented by the black spot.

11.3 Surface Charge Densities

The surface charge densities inside and outside the hollow shell (that is, along the internal and external surfaces), σ_i and σ_o, are obtained utilizing Gauss's law. They have the same value and are given by

$$\sigma_i = \sigma_o = \frac{\varepsilon_0 \phi_B}{2\pi a} \tan\frac{\varphi}{2} = \frac{\varepsilon_0 \phi_B}{2\pi a} \tan\psi \,. \tag{11.23}$$

A plot of this surface charge density as a function of φ is shown in Figure 11.4. The total charge density σ_t is given by

$$\sigma_t = \sigma_i + \sigma_o = \frac{\varepsilon_0 \phi_B}{\pi a} \tan\frac{\varphi}{2} = \frac{\varepsilon_0 \phi_B}{\pi a} \tan\psi \,. \tag{11.24}$$

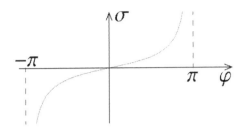

Figure 11.4: Surface charge densities $\sigma \equiv \sigma_i = \sigma_o$ inside and outside the hollow shell as a function of the azimuthal angle φ, according to Eq. (11.23) [226].

Expanding Eq. (11.23) for $\varphi \ll 1$ rad and utilizing Eqs. (11.1) and (11.11) yields:

$$\sigma_i(\varphi \ll 1 \text{ rad}) = \sigma_o(\varphi \ll 1 \text{ rad}) \approx \frac{\varepsilon_0 \phi_B \varphi}{4\pi a} = \frac{\varepsilon_0 \phi}{2a} \ . \qquad (11.25)$$

This result coincides with Eq. (9.7) when we are over the plate ($x = 0$) and $\sigma_A = 0$, if we equate the circumference of the cylindrical shell here ($2\pi a$) with the longitudinal length ℓ_z of the plate of Section 9.2, as expected. The surface charge density which appears in Eq. (9.7) is the total charge density due to the charges in both sides of the plate, analogous to $\sigma_i + \sigma_o$ of Eq. (11.25).

On the other hand, for $\varphi = (\pi \pm \delta)$ rad, with $0 < \delta \ll 1$ (that is, close to $\varphi = \pi$ rad) we have:

$$\sigma_i = \sigma_o \approx \mp \frac{\varepsilon_0 \phi_B}{\pi a \delta} = \mp \frac{\varepsilon_0 \phi_B}{\pi s} \ . \qquad (11.26)$$

Here $s \equiv a\delta$ is the distance along the surface of the cylindrical shell to the line battery. This is an important result which shows that the surface charge density diverges inversely proportional to the distance from the line battery in this idealized case.

Eq. (11.23) indicates that in regions close to the battery the surface charge density is no longer a linear function of the longitudinal coordinate (in this case the arc $a\varphi$) of the resistive conductor. It is linear only close to $\varphi = 0$ rad but increases nonlinearly (that is, it is not proportional to $a\varphi$) toward the battery. See Figure 11.4. This nonlinearity should also occur in straight conductors when we are close to the battery. This has been confirmed in 2004 and 2005 [205, 222].

Eqs. (11.14) to (11.26) indicate that several functions are proportional to the emf of the battery, namely: the internal and external potential and electric field, as well as the surface charge densities in the internal and external walls. That is, they are proportional to the voltage $\phi(\pi) - \phi(-\pi) = \phi_B$ between the terminals of the battery. Suppose we have two cylindrical shells 1 and 2 of the same radius but with different resistivities. If we connect only shell 1 with battery B and later on if we connect only shell 2 with the same battery B (assuming the battery has not lost its power), different steady currents will flow in each shell, as they have different resistivities. But the internal and external potentials, electric fields and surface charge densities will be the same in both cases. This again illustrates that the electric field outside a resistive conductor carrying a steady current is proportional to the voltage to which it is subjected. The importance of the present case is that this has been shown in a situation in which we were able to find an exact analytical solution of all magnitudes. That is, this external electric field does not depend directly upon the current flowing in the circuit. After all, the electric field was found to be the same even when different currents flow in two circuits connected by the same emf. In order to observe the effects of the external electric field, it is most important to work with circuits connected to high voltage sources, as this field is proportional to the applied emf.

11.4 Representation in Fourier Series

Our solution of the potential in terms of Fourier series was presented in Eqs. (11.12) and (11.13). These series can be put in closed form. See Eqs. (11.14) and (11.15). If this were not possible, we could continue to utilize the Fourier series representation obtaining the electric field $\vec{E} = -\nabla \phi$ in the form

$$\vec{E}(\rho < a, \varphi, z) = \frac{\phi_B}{\pi \rho} \left\{ \sum_{m=1}^{\infty} \left(\frac{-\rho}{a} \right)^m [\sin(m\varphi)\hat{\rho} + \cos(m\varphi)\hat{\varphi}] \right\}, \qquad (11.27)$$

$$\vec{E}(\rho > a, \varphi, z) = -\frac{\phi_B}{\pi \rho} \left\{ \sum_{m=1}^{\infty} \left(\frac{-a}{\rho} \right)^m [\sin(m\varphi)\hat{\rho} - \cos(m\varphi)\hat{\varphi}] \right\}. \qquad (11.28)$$

From these two equations the tangential component of the electric field at $\rho = a$ is given by:

$$E_\varphi(a, \varphi, z) = \frac{\phi_B}{\pi a} \left[\sum_{m=1}^{\infty} (-1)^m \cos(m\varphi) \right]. \qquad (11.29)$$

This is a divergent series. This happens with the differentiation of some Fourier series [202, Section 14.4]. By differentiating both sides of Eq. (11.11) we obtain

$$1 = 2 \left[\sum_{m=1}^{\infty} (-1)^{m-1} \cos(n\varphi) \right]. \qquad (11.30)$$

While Eq. (11.11) is convergent, the latter series is divergent. But if we disregard this and apply Eq. (11.30) into Eq. (11.29) we obtain the same result as Eq. (11.22). And this is a reasonable result.

Figure 11.5 is a plot of

$$f(\varphi) \equiv \sum_{m=1}^{\infty} (-1)^{m-1} \cos(n\varphi) , \qquad (11.31)$$

including 100 terms in the summation. The oscillations are due to the divergent character of this series representation, although we can see that the curve oscillates around the constant value of $f(\varphi) = 1/2$, which was expected according to Eq. (11.30).

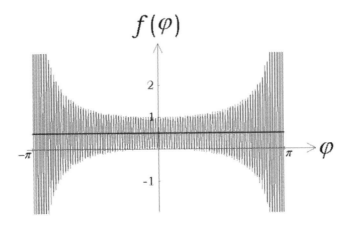

Figure 11.5: Plot of Eq. (11.31) including 100 terms in the summation. The oscillations are due to the divergent character of this series. The bold line is a plot of the constant 1/2 as expected by Eq. (11.30).

In order to deal with a divergent Fourier series, we thought of applying an average approach. In particular, at each angle φ_i we consider the average value of a generic function $g(\varphi)$, $\overline{g(\varphi_i)}$, namely:

$$\overline{g(\varphi_i)} \equiv \frac{1}{\Delta\varphi} \int_{\varphi_i-\Delta\varphi/2}^{\varphi_i+\Delta\varphi/2} g(\varphi)d\varphi . \qquad (11.32)$$

The value of $\Delta\varphi$ is typically taken as the whole interval in which we are plotting $g(\varphi)$ divided by the number of terms we are including in the summation. For instance, if we are plotting a function $g(\varphi)$ in the interval $\varphi = -\pi$ rad to $\varphi = \pi$ rad and we include 100 terms in the Fourier series expansion of $g(\varphi)$, then $\Delta\varphi = (2\pi/100)$ rad.

Figure 11.6 is a plot of $\overline{f(\varphi)}$ obtained from this averaging approach utilizing Eq. (11.31). From this Figure we can see that $\overline{f(\varphi)}$ coincides with the expected value of 1/2, indicating the correctness of this averaging procedure.

The surface charge densities inside and outside the shell can be obtained from Gauss's law. From Eqs. (11.27) and (11.28) this yields:

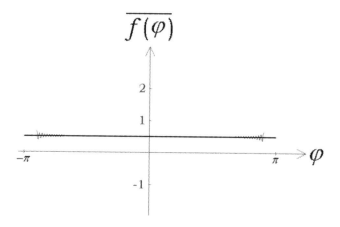

Figure 11.6: Plot of $\overline{f(\varphi)}$ obtained from Eqs. (11.32) and (11.31), overlaid on the constant value $1/2$. As $\overline{f(\varphi)}$ coincides with this constant value, this indicates the correctness of this averaging procedure.

$$\sigma_i = -\lim_{\rho \to a} \varepsilon_0 \vec{E}(\rho < a) \cdot \hat{\rho} = -\frac{\varepsilon_0 \phi_B}{\pi a} \left[\sum_{m=1}^{\infty} (-1)^m \sin(m\varphi) \right], \quad (11.33)$$

$$\sigma_o = \lim_{\rho \to a} \varepsilon_0 \vec{E}(\rho > a) \cdot \hat{\rho} = \sigma_i. \quad (11.34)$$

The same expressions are obtained from Eq. (11.23) by expanding $\tan(\varphi/2)$ in Fourier series.

The total charge density expressed in Fourier series is given by

$$\sigma_t = \sigma_i + \sigma_o = -\frac{2\varepsilon_0 \phi_B}{\pi a} \left[\sum_{m=1}^{\infty} (-1)^m \sin(m\varphi) \right]. \quad (11.35)$$

In Figure 11.7 we present a plot of Eq. (11.35) including 100 terms in the summation. The oscillations in this Figure are probably due to convergence problems of the Fourier series already discussed. Increasing the number of terms does not improve significantly the plot or decrease the amplitude of oscillation around each value of φ. The bold line in this Figure is given by Eq. (11.24).

Fig. 11.8 is a plot of Eq. (11.24) overlaid on a plot of $\overline{\sigma_t(\varphi_i)}$ obtained from Eqs. (11.35) and (11.32). In this case we have considered a whole oscillation of $\sigma_t(\varphi_i)$ around each angle φ_i. The two plots coincided with one another (the two curves are indistinguishable in Fig. 11.8), indicating the correctness of this averaging procedure.

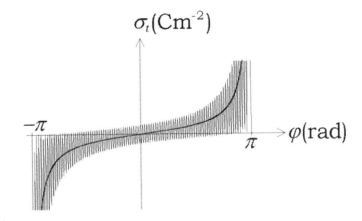

Figure 11.7: Total surface charge density $\sigma_t = \sigma_i + \sigma_o$ of an infinite resistive cylindrical shell carrying a steady azimuthal current as a function of the angle φ. The bold line is a plot of the closed form solution of $\sigma_t(\varphi)$, Eq. (11.24), while the oscillatory line is a plot of $\sigma_t(\varphi)$ expressed in a Fourier series, Eq. (11.35).

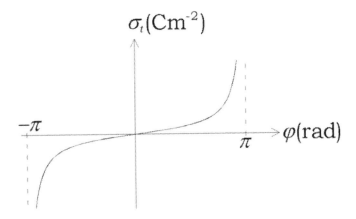

Figure 11.8: Total surface charge density σ_t. The summation that appears as an oscillation in Fig. 11.7, given by Eq. (11.35), is smoothed out by taking the mean value for each point of its surroundings (in this case, a whole oscillation around each point), utilizing Eq. (11.32). The closed analytical form, Eq. (11.24), is overlaid on it. Both plots coincide with one another, indicating the correctness of this averaging procedure.

11.5 Lumped Resistor

Heald also considered a lumped resistor, *i.e.*, a cylindrical shell of finite resistivity for $-\alpha < \varphi < \alpha$ and zero resistivity outside this region [226]. The potential at the shell was given by

$$\phi(a,\ \varphi,\ z) = \sum_{k=1}^{\infty} A_k \sin(k\varphi)\ . \qquad (11.36)$$

Here the coefficients A_k are given by:

$$A_k = \frac{\phi_B}{\pi}\left[\int_0^\alpha \frac{\varphi}{\alpha}\sin(k\varphi)d\varphi + \int_\alpha^\pi \sin(k\varphi)d\varphi\right]$$

$$= \frac{\phi_B}{\pi}\left[\frac{(-1)^{k-1}}{k} + \frac{\sin(k\varphi)}{k^2\alpha}\right]\ . \qquad (11.37)$$

The potential inside and outside the shell is given by, respectively:

$$\phi(\rho \leq a,\ \varphi,\ z) = \frac{\phi_B}{\pi}\left[\tan^{-1}\frac{\rho\sin\varphi}{a+\rho\cos\varphi}\right.$$

$$\left. + \sum_{k=1}^{\infty}\frac{\rho^k}{a}\frac{\sin(k\alpha)}{k^2\alpha}\sin(k\varphi)\right]\ , \qquad (11.38)$$

$$\phi(\rho \geq a,\ \varphi,\ z) = \frac{\phi_B}{\pi}\left[\tan^{-1}\frac{a\sin\varphi}{\rho+a\cos\varphi}\right.$$

$$\left. + \sum_{k=1}^{\infty}\frac{a^k}{\rho}\frac{\sin(k\alpha)}{k^2\alpha}\sin(k\varphi)\right]\ . \qquad (11.39)$$

The equipotentials for this case of lumped resistor are represented in Figure 11.9.

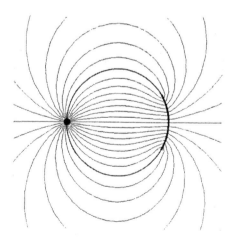

Figure 11.9: Equipotential lines for the lumped resistor.

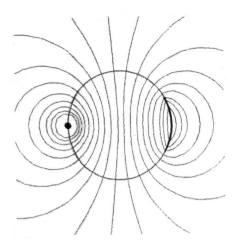

Figure 11.10: Lines of electric field for the lumped resistor.

The electric field lines are represented in Figure 11.10.

The internal and external surface charge densities are again equal. In this case they are given by:

$$\sigma_i = \sigma_o = \frac{\varepsilon_0 \phi_B}{2\pi a} \left[\tan \frac{\varphi}{2} + \sum_{k=1}^{\infty} 2 \frac{\sin(k\alpha)}{k\alpha} \sin(k\varphi) \right]$$

$$= \frac{\varepsilon_0 \phi_B}{2\pi a} \left[\tan \frac{\varphi}{2} + \frac{1}{\alpha} \ln \left| \frac{\sin[(\varphi + \alpha)/2]}{\sin[(\varphi - \alpha)/2]} \right| \right] . \quad (11.40)$$

In this case the surface charge densities diverge not only at the battery but also at the discontinuity in the resistivity of the shell, as in Figure 11.11.

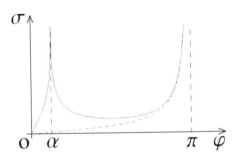

Figure 11.11: Densities of surface charge $\sigma \equiv \sigma_i = \sigma_o$ along the internal and external surfaces of the hollow lumped resistor (continuous line) as a function of the azimuthal angle, as given by Eq. (11.40). The dashed line represents the previous case of a uniformly resistive conductor.

Another qualitative discussion of lumped resistors can be found in the book of Chabay and Sherwood [166, Section 18.6]. Their analysis is extremely didactic and helpful.

Chapter 12

Resistive Spherical Shell with Azimuthal Current

12.1 Introduction

Our goal in this chapter is to consider a steady azimuthal current flowing in a resistive spherical shell [227]. The mathematical difficulty is intermediate between the infinite cylindrical shell which we considered in the previous Chapter and the toroidal conductor which is the subject of the next Chapter. The importance of the present case is that we can obtain exact analytical solutions for the external and internal distribution of surface charges, potential and electric field which are not as complex as in the toroidal conductor. Despite this fact they show clearly the existence of an electric field outside a resistive conductor bounded in a finite volume of space. To the best of our knowledge this case has never been treated before by other authors.

12.2 Description of the Problem

Consider a resistive spherical shell of radius a, centered at the origin. We suppose an idealized linear battery located along a meridian of the shell (like Greenwich Meridian) and maintaining a constant potential difference between its left and right sides. See Figure 12.1.

That is, the battery is a semi-circumference in the plane $y = 0$ with its extremities at $(x, y, z) = (0, 0, \pm a)$ and central point along the semi-circumference at $(x, y, z) = (-a, 0, 0)$. Utilizing spherical coordinates (r, θ, φ) the linear battery is then located at (a, θ, π). We suppose that the potential difference generated by the battery does not depend upon the polar angle θ. The battery generates a steady current flowing along the shell in the azimuthal direction $-\hat{\varphi}$. See Figs. 12.1 and 12.2. The medium inside and outside the spherical shell is supposed to be air or vacuum.

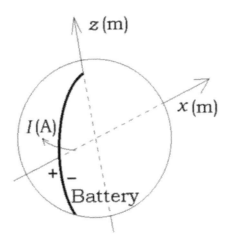

Figure 12.1: A resistive spherical shell of radius a (m) is centered at the origin. An idealized linear battery located at $(r, \theta, \varphi) = (a, \theta, \pi)$ generates a steady current I (A) flowing along the surface of the shell in the azimuthal direction $-\hat{\varphi}$. The bold line represents the battery, which has the form of a semi-circumference.

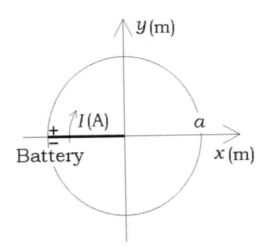

Figure 12.2: Projection of the resistive spherical shell with radius a in the plane $z = 0$. Notice that the battery, represented by the bold line, is seen as a straight line for $-a \leq x \leq 0$. In this plane the current flows in the clockwise direction $-\hat{\varphi}$.

According to Ohm's law, the potential ϕ along the surface is given by (including also a constant potential for the sake of generality, so that we can return to the situation of a charged shell without current as a special case):

$$\phi(a, \theta, \varphi) = \phi_A + \phi_B \frac{\varphi}{2\pi} \,. \tag{12.1}$$

The goal is to find solutions of Laplace's equation $\nabla^2 \phi = 0$ outside and inside the spherical shell utilizing Eq. (12.1) as a boundary condition, together with finite values of the potential at the center of the shell and at infinity. The electric field is then found by $\vec{E} = -\nabla \phi$. Lastly the surface charge density σ is obtained by the standard procedure of taking the radial components of the external and internal electric fields when $r \to a$.

12.3 General Solution

Laplace's equation in spherical coordinates can be written as:

$$\nabla^2 \phi = \frac{\partial^2 \phi}{\partial r^2} + \frac{2}{r} \frac{\partial \phi}{\partial r} + \frac{1}{r^2} \frac{\partial^2 \phi}{\partial \theta^2} + \frac{\cot \theta}{r^2} \frac{\partial \phi}{\partial \theta} + \frac{1}{r^2 \sin^2 \theta} \frac{\partial^2 \phi}{\partial \varphi^2} = 0 \,. \tag{12.2}$$

The electric potential ϕ can be solved utilizing the method of separation of variables, $\phi(r, \theta, \varphi) = R(r)\Theta(\theta)\Phi(\varphi)$. This yields the following equations for the functions R, Θ and Φ [224, pp. 24–27]:

$$R'' + \frac{2}{r} R' - \frac{\alpha_2}{r^2} R = 0 \,, \tag{12.3}$$

$$\Theta'' + \Theta' \cot \theta + \left(\alpha_2 - \frac{\alpha_1}{\sin^2 \theta} \right) \Theta = 0 \,, \tag{12.4}$$

$$\Phi'' + \alpha_1 \Phi = 0 \,, \tag{12.5}$$

where α_1 and α_2 are constants. The function $\Phi(\varphi)$ must be periodic in φ, i.e., $\Phi(0) = \Phi(2\pi)$. This implies $\alpha_1 = q^2$, where $q = 0, 1, 2, \ldots$ The solutions of Eq. (12.5) are then $\Phi_q^{(1)} = \sin(q\varphi)$ and $\Phi_q^{(2)} = \cos(q\varphi)$. Eq. (12.4) is the associated Legendre equation [202, Sec. 12.5]. In order to have finite solutions at $\theta = 0$ rad and at $\theta = \pi$ rad the constant α_2 must have the form $\alpha_2 = p(p+1)$, with $p = 0, 1, 2, \ldots$ The solutions of Eq. (12.4) are then the associated Legendre functions of first and second kind, namely, $\Theta_{pq}^{(1)} = P_p^q(\cos \theta)$ and $\Theta_{pq}^{(2)} = Q_p^q(\cos \theta)$. When $q = 0$ they reduce to Legendre polynomial, $P_p(\cos \theta)$, and to Legendre function of the second kind, $Q_p(\cos \theta)$, respectively. The solutions of Eq. (12.3) with $\alpha_2 = p(p+1)$ are given by $R_p^{(1)} = r^p$ and $R_p^{(2)} = r^{-p-1}$.

The potential must remain finite at every point in space. The solution $R_p^{(1)} = r^p$ diverges when $r \to \infty$ and $p \geq 1$. For this reason we eliminate this solution outside the shell. By specifying that the potential goes to zero when $r \to \infty$ we can also eliminate the solution with $p = 0$. Analogously we eliminate the solution $R_p^{(2)} = r^{-p-1}$ inside the shell as it diverges when $r \to 0$. The function $P_p^q(\cos \theta)$ is finite for 0 rad $\leq \theta \leq \pi$ rad. On the other hand, $Q_p^q(\cos \theta)$ diverges at $\theta = 0$ rad and at $\theta = \pi$ rad. We then eliminate this solution both inside and outside the shell. The finite solutions for the potential outside and inside the shell are then given by the combination of all possible values of $R_p(r)$, $\Theta_{pq}(\theta)$ and $\Phi_q(\varphi)$, respectively:

$$\phi_o(r \geq a, \theta, \varphi) = \sum_{p=0}^{\infty} r^{-(p+1)} \Big\{ A_p P_p(\cos\theta)$$

$$+ \sum_{q=1}^{\infty} [B_{pq} \sin(q\varphi) + C_{pq} \cos(q\varphi)] P_p^q(\cos\theta) \Big\} , \quad (12.6)$$

$$\phi_i(r \leq a, \theta, \varphi) = \sum_{p=0}^{\infty} r^p \Big\{ D_p P_p(\cos\theta)$$

$$+ \sum_{q=1}^{\infty} [E_{pq} \sin(q\varphi) + F_{pq} \cos(q\varphi)] P_p^q(\cos\theta) \Big\} . \quad (12.7)$$

In order to obtain the coefficients A_p, B_{pq}, C_{pq}, D_p, E_{pq} and F_{pq} we must apply the boundary condition at the surface of the shell, $r = a$. Expanding Eq. (12.1) in Fourier series [226]:

$$\phi(a, \theta, \varphi) = \phi_A + \phi_B \frac{\varphi}{2\pi} = \phi_A + \frac{\phi_B}{\pi} \left[\sum_{q=1}^{\infty} \frac{(-1)^{q-1}}{q} \sin(q\varphi) \right] . \quad (12.8)$$

As there are no terms in $\cos(q\varphi)$ in Eq. (12.8) we obtain immediately $C_{pq} = F_{pq} = 0$.

First we find the coefficients A_p and B_{pq} for the region outside the shell ($r \geq a$). Eq. (12.6) calculated at $r = a$ combined with Eq. (12.8) yields the following equations:

$$\phi_A = \sum_{p=0}^{\infty} a^{-(p+1)} A_p P_p(\cos\theta) , \quad (12.9)$$

$$\frac{\phi_B}{\pi} \frac{(-1)^{q-1}}{q} = \sum_{p=0}^{\infty} a^{-(p+1)} B_{pq} P_p^q(\cos\theta) . \quad (12.10)$$

To find the coefficients A_p and B_{pq} we multiply both sides of Eq. (12.9) by $P_\ell(\cos\theta) \sin\theta \, d\theta$, both sides of Eq. (12.10) by $P_\ell^q(\cos\theta) \sin\theta \, d\theta$, and integrate from 0 rad to π rad. We then utilize the orthogonality relation of Legendre polynomials [202, Eq. (12.104)]:

$$\int_0^\pi P_p^q(\cos\theta) P_\ell^q(\cos\theta) \sin\theta \, d\theta = \frac{2}{2p+1} \frac{(p+q)!}{(p-q)!} \delta_{p\ell} , \quad (12.11)$$

where $\delta_{p\ell}$ is Kronecker's delta function, which is 1 for $p = q$ and 0 for $p \neq q$. This yields:

$$A_p = a\phi_A \delta_{p0} , \quad (12.12)$$

and
$$B_{pq} = \frac{\phi_B}{\pi} a^{p+1} \frac{(-1)^{q-1}}{q} \frac{2p+1}{2} \frac{(p-q)!}{(p+q)!} I_{pq}, \qquad (12.13)$$

where we defined:
$$I_{pq} \equiv \int_0^\pi P_p^q(\cos\theta) \sin\theta \, d\theta . \qquad (12.14)$$

Notice that $I_{pq} = 0$ for $p+q$ odd due to the parity property of the associated Legendre functions [202, p. 725].

We can change the upper limit of the summation over q in Eq. (12.6) from ∞ to p, because $P_p^q(\xi) = 0$ for $q > p$. The final solution for the potential outside a spherical shell conducting a steady azimuthal current is given by:

$$\phi_o(r \geq a, \theta, \varphi) = \phi_A \frac{a}{r} + \frac{\phi_B}{2\pi} \left[\sum_{p=1}^{\infty} \sum_{q=1}^{p} \frac{a^{p+1}}{r^{p+1}} \frac{(-1)^{q-1}}{q} (2p+1) \frac{(p-q)!}{(p+q)!} \times \right.$$

$$\left. \times I_{pq} P_p^q(\cos\theta) \sin(q\varphi) \right] . \qquad (12.15)$$

It is useful to keep in mind that the summation order can be inverted, from $\sum_{p=1}^{\infty} \sum_{q=1}^{p}$ to $\sum_{q=1}^{\infty} \sum_{p=q}^{\infty}$.

For the region far from the origin, $r \gg a$, the two most relevant terms of Eq. (12.15) are:

$$\phi_o(r \gg a, \theta, \varphi) \approx \phi_A \frac{a}{r} + \phi_B \frac{3a^2}{8r^2} \sin\theta \sin\varphi . \qquad (12.16)$$

This can be understood as the potential of a point charge $q_{\text{sphere}} = 4\pi\varepsilon_0 \phi_A a$ at the center of the shell plus the potential of an electric dipole of moment $\vec{p}_{\text{sphere}}$ located at the origin with $\vec{p}_{\text{sphere}} = (3\pi\varepsilon_0 \phi_B a^2/2)\hat{y}$, namely:

$$\phi_o(r \gg a, \theta, \varphi) = \frac{q_{\text{sphere}}}{4\pi\varepsilon_0 r} + \frac{\vec{p}_{\text{sphere}} \cdot \vec{r}}{4\pi\varepsilon_0 r^3} . \qquad (12.17)$$

The solution for the potential inside the sphere ($r \leq a$), ϕ_i, can be found by changing $(a/r)^{p+1} \to (r/a)^p$, as discussed by Jackson [13, p. 101]:

$$\phi_i(r \leq a, \theta, \varphi) = \phi_A + \frac{\phi_B}{2\pi} \left[\sum_{p=1}^{\infty} \sum_{q=1}^{p} \frac{r^p}{a^p} \frac{(-1)^{q-1}}{q} (2p+1) \frac{(p-q)!}{(p+q)!} \times \right.$$

$$\left. \times I_{pq} P_p^q(\cos\theta) \sin(q\varphi) \right] , \qquad (12.18)$$

where I_{pq} is given by Eq. (12.14).

Utilizing that (as can be seen multiplying both sides of Eq. (12.19) by $P_\ell^q(\cos\theta)\sin\theta\, d\theta$, integrating from $\theta = 0$ rad to $\theta = \pi$ rad and finally applying Eqs. (12.14), (12.13) and (12.8)):

$$\sum_{p=1}^{\infty}\sum_{q=1}^{p}\frac{(-1)^{q-1}}{q}(2p+1)\frac{(p-q)!}{(p+q)!}I_{pq}P_p^q(\cos\theta)\sin(q\varphi) = \varphi, \qquad (12.19)$$

we obtain from Eqs. (12.15) and (12.18) in the limit $r \to a$ that $\phi_o(a,\theta,\varphi) = \phi_i(a,\theta,\varphi) = \phi_A + \phi_B\varphi/2\pi$, as expected.

12.4 Electric Field and Surface Charges

The electric field in spherical coordinates is given by:

$$\vec{E} = -\nabla\phi = -\frac{\partial\phi}{\partial r}\hat{r} - \frac{1}{r}\frac{\partial\phi}{\partial\theta}\hat{\theta} - \frac{1}{r\sin\theta}\frac{\partial\phi}{\partial\varphi}\hat{\varphi}. \qquad (12.20)$$

This yields the following components outside and inside the shell, respectively:

$$E_{r,o} = \phi_A\frac{a}{r^2} + \frac{\phi_B}{2\pi}\left[\sum_{p=1}^{\infty}\sum_{q=1}^{p}\frac{a^{p+1}}{r^{p+2}}\frac{(-1)^{q-1}}{q}(p+1)(2p+1)\frac{(p-q)!}{(p+q)!}\times\right.$$
$$\left.\times I_{pq}P_p^q(\cos\theta)\sin(q\varphi)\right], \qquad (12.21)$$

$$E_{\theta,o} = \frac{\phi_B}{2\pi}\left[\sum_{p=1}^{\infty}\sum_{q=1}^{p}\frac{a^{p+1}}{r^{p+2}}\frac{(-1)^{q-1}}{q}(2p+1)\frac{(p-q)!}{(p+q)!}I_{pq}P_p^{q\prime}(\cos\theta)\sin\theta\sin(q\varphi)\right], \qquad (12.22)$$

$$E_{\varphi,o} = -\frac{\phi_B}{2\pi}\left[\sum_{p=1}^{\infty}\sum_{q=1}^{p}\frac{a^{p+1}}{r^{p+2}}(-1)^{q-1}(2p+1)\frac{(p-q)!}{(p+q)!}I_{pq}\frac{P_p^q(\cos\theta)}{\sin\theta}\cos(q\varphi)\right], \qquad (12.23)$$

$$E_{r,i} = -\frac{\phi_B}{2\pi}\left[\sum_{p=1}^{\infty}\sum_{q=1}^{p}\frac{r^{p-1}}{a^p}\frac{(-1)^{q-1}}{q}p(2p+1)\frac{(p-q)!}{(p+q)!}I_{pq}P_p^q(\cos\theta)\sin(q\varphi)\right], \qquad (12.24)$$

$$E_{\theta,i} = \frac{\phi_B}{2\pi}\left[\sum_{p=1}^{\infty}\sum_{q=1}^{p}\frac{r^{p-1}}{a^p}\frac{(-1)^{q-1}}{q}(2p+1)\frac{(p-q)!}{(p+q)!}I_{pq}P_p^{q\prime}(\cos\theta)\sin\theta\sin(q\varphi)\right], \qquad (12.25)$$

$$E_{\varphi,i} = -\frac{\phi_B}{2\pi}\left[\sum_{p=1}^{\infty}\sum_{q=1}^{p}\frac{r^{p-1}}{a^p}(-1)^{q-1}(2p+1)\frac{(p-q)!}{(p+q)!}I_{pq}\frac{P_p^q(\cos\theta)}{\sin\theta}\cos(q\varphi)\right]. \qquad (12.26)$$

In Eqs. (12.22) and (12.25) $P_p^{q\prime}(\xi)$ is the derivative of the associated Legendre function $P_p^q(\xi)$ relative to its argument ξ.

From Eqs. (12.22), (12.25) and (12.19) we obtain in the limit $r \to a$ that:

$$E_{\theta,o}(a,\theta,\varphi) = E_{\theta,i}(a,\theta,\varphi)$$

$$= \frac{\phi_B}{2\pi a}\left[\sum_{p=1}^{\infty}\sum_{q=1}^{p}\frac{(-1)^{q-1}}{q}(2p+1)\frac{(p-q)!}{(p+q)!}I_{pq}P_p^{q'}(\cos\theta)\sin\theta\sin(q\varphi)\right]$$

$$= \frac{\phi_B}{2\pi a}\frac{d}{d\theta}\left[\sum_{p=1}^{\infty}\sum_{q=1}^{p}\frac{(-1)^{q-1}}{q}(2p+1)\frac{(p-q)!}{(p+q)!}I_{pq}P_p^{q}(\cos\theta)\sin(q\varphi)\right]$$

$$= \frac{\phi_B}{2\pi a}\frac{d}{d\theta}\varphi = 0 \ . \tag{12.27}$$

From Eqs. (12.23), (12.26) and (12.19) we obtain in the limit $r = a$ that:

$$E_{\varphi,o}(a,\theta,\varphi) = E_{\varphi,i}(a,\theta,\varphi)$$

$$= -\frac{\phi_B}{2\pi a\sin\theta}\left[\sum_{p=1}^{\infty}\sum_{q=1}^{p}(-1)^{q-1}(2p+1)\frac{(p-q)!}{(p+q)!}I_{pq}P_p^{q}(\cos\theta)\cos(q\varphi)\right]$$

$$= -\frac{\phi_B}{2\pi a\sin\theta}\frac{d}{d\varphi}\left[\sum_{p=1}^{\infty}\sum_{q=1}^{p}\frac{(-1)^{q-1}}{q}(2p+1)\frac{(p-q)!}{(p+q)!}I_{pq}P_p^{q}(\cos\theta)\sin(q\varphi)\right]$$

$$= -\frac{\phi_B}{2\pi a\sin\theta}\frac{d}{d\varphi}\varphi = -\frac{\phi_B}{2\pi a\sin\theta} \ . \tag{12.28}$$

Eq. (12.27) indicates that the non-radial electric field at the surface of the shell is only in the azimuthal direction, as expected from Eq. (12.1). The length of an azimuthal circumference at the polar angle θ along the surface of the shell is given by $2\pi a\sin\theta$. Eq. (12.28) indicates that $E_\varphi(a,\theta,\varphi)$ at each polar angle θ is given by the total electromotive force, $\Delta\phi = \phi_B$, over the length of the corresponding circuit at the polar angle θ, as expected. By Ohm's law the same inverse proportionality with $\sin\theta$ will be valid for the surface current density. That is, $\vec{K}$ should be proportional to $\phi_B/\sin\theta$. According to this model the current density should diverge at the poles (in $\theta = 0$ rad and in $\theta = \pi$ rad). This indicates a limitation for the theoretical model which we are utilizing. This divergence arises due to the fact that we are utilizing a conducting spherical shell with an idealized linear battery along a meridian. In a real experiment this divergence should not occur. This means that our analytical theoretical solution obtained in this Section should not be valid close to these two poles when compared with a real experiment. The reason for utilizing our theoretical model is that it yields an analytical solution for the important problem of a closed current flowing in a finite volume of space.

The surface charge distributions outside and inside the shell are related to the electric field through Gauss's law:

$$\sigma_o(a,\theta,\varphi) = \lim_{r\to a}\varepsilon_0\vec{E}_o(r,\theta,\varphi)\cdot\hat{r} = \varepsilon_0\left\{\frac{\phi_A}{a} + \frac{\phi_B}{2\pi a}\sum_{p=1}^{\infty}\left[\sum_{q=1}^{p}\frac{(-1)^{q-1}}{q}\times\right.\right.$$

$$\times (p+1)(2p+1)\frac{(p-q)!}{(p+q)!}I_{pq}P_p^q(\cos\theta)\sin(q\varphi)\Bigg]\Bigg\}\;, \qquad (12.29)$$

$$\sigma_i(a,\theta,\varphi) = -\lim_{r\to a}\varepsilon_0\vec{E}_i(r,\theta,\varphi)\cdot\hat{r}$$

$$= \varepsilon_0\frac{\phi_B}{2\pi a}\Bigg[\sum_{p=1}^{\infty}\sum_{q=1}^{p}\frac{(-1)^{q-1}}{q}p(2p+1)\frac{(p-q)!}{(p+q)!}I_{pq}P_p^q(\cos\theta)\sin(q\varphi)\Bigg]\;. \qquad (12.30)$$

In this case of a spherical shell we have $\sigma_o(a,\theta,\varphi)\neq\sigma_i(a,\theta,\varphi)$. In the cylindrical case, on the other hand, we obtained the same surface charge densities both inside and outside the shell. The total surface charge density is the sum of these two expressions, namely, $\sigma_t=\sigma_o+\sigma_i$.

In Fig. 12.3 we plot the equipotentials in the plane $z=0$ of the spherical shell with $\phi_A=0$ (no net charge in the shell). The current is in the clockwise direction, the bold circumference represents the shell. The electric field lines which are perpendicular to these equipotentials are also contained in the plane $z=0$. This can be seen noting that for $\theta=\pi/2$ rad we have $P_p^{q\prime}(\cos\theta)=0$ for $p+q$ even (see page 733 of the book by Arfken and Weber [202] combined with the recurrence relation (12.87) of the same work). Using the property that I_{pq} is null for $p+q$ odd, we have that $E_\theta=0$ for both $r<a$ and $r>a$.

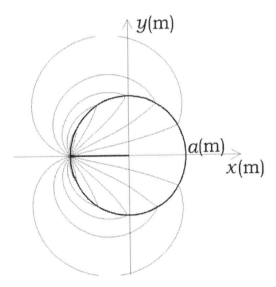

Figure 12.3: Equipotentials in the plane $z=0$. The resistive spherical shell carries a clockwise steady current. The bold circumference represents the shell. The projection of the battery is represented by the bold straight line going from $x=-a$ to $x=0$. The electric field has no z component, so the electric field lines are orthogonal to the equipotentials in this plane.

In Fig. 12.4 we plot the equipotentials in the plane $x=0$. The current enters the plane of the paper on the left side of the bold circumference and leaves the plane of the paper on the right side. We utilized $\phi_A = 0$. In this case the electric field lines are not contained in this plane (E_φ or E_x are not null in the entire plane).

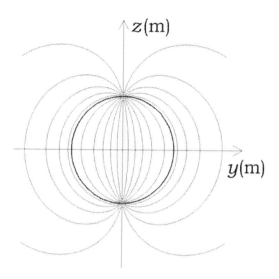

Figure 12.4: Equipotentials in the plane $x = 0$ of the spherical shell with $\phi_A = 0$. The bold circumference represents the shell. The current enters the plane of the paper on the left side of the circumference and leaves the paper on the right side.

In Fig. 12.5 we plot the total surface charge density σ_t in the equatorial plane as a function of the azimuthal angle φ, normalized by the value of σ_t at $\varphi = \pi/4$ rad. The presence of the term $(-1)^q \sin(q\varphi)$ in Eqs. (12.29) and (12.30) causes a rapid variation in the calculation of σ_t. This can be seen in the oscillation of Fig. 12.5.

The oscillations on the plot of $\sigma_t(\varphi)$ shown in Figs. 12.5 and 11.7 probably occur because σ_t is proportional to the radial component of the electric field that comes from differentiating a Fourier series. And sometimes there are convergence problems with the differentiation of Fourier series, as we saw in Section 11.4. By raising the number of terms in the Fourier series of σ_t we increase only the number of oscillations in the curves.

We did not succeed in putting the series solutions given by Eqs. (12.29) and (12.30) in closed analytical form. But utilizing the averaging procedure presented in Section 11.4, we obtained Fig. 12.6. The wiggles around $\varphi = \pm\pi/2$ rad should be due to numerical approximations without physical significance. The real curve should be smooth like Fig. 11.8. Fig. 12.6 indicates that $\sigma_t(\varphi)$ is linear with φ far from the battery (i.e., around $\varphi = 0$ rad), diverging close to it (when $\varphi \to \pm\pi$ rad). This is the important physical result.

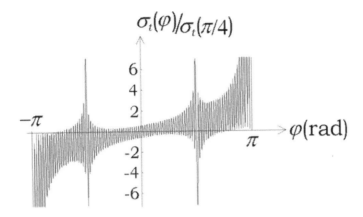

Figure 12.5: Total surface charge density $\sigma_t(\varphi)$ as a function of the azimuthal angle φ in the equatorial plane $z = 0$ of a resistive spherical shell carrying a steady azimuthal current, normalized by its value at $\varphi = \pi/4$ rad. We have utilized Eqs. (12.29) and (12.30) with the summation in p going from $p = 1$ to $p = 100$.

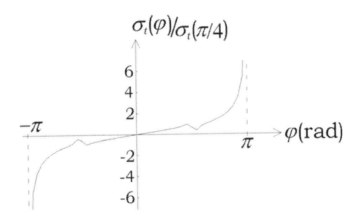

Figure 12.6: Smoothed out plot of Fig. 12.5.

12.5 Conclusion

We have obtained the surface charge density, σ, potential, ϕ, and electric field, $\vec{E}$, outside and inside a resistive spherical shell carrying a steady azimuthal current. We have plotted the total surface charge density σ_t as a function of the azimuthal angle φ. We have found that σ_t is linear with φ far from the battery, diverging to infinity close to it. At great distances from the spherical shell the potential is that of a point charge plus that of an electric dipole, Eq. (12.17). The total charge q and dipole moment $\vec{p}$ of this system are given by Eq. (12.17) and in the paragraph before it. Alternatively, they can also be found by $q = \int\int \sigma \, da$ and $\vec{p} = \int\int \sigma \vec{r} \, da$, where da is an area element and the integration is over the surface of the system. The two approaches agree with one another, as expected.

Chapter 13

Resistive Toroidal Conductor with Azimuthal Current

13.1 Introduction

The calculations of this Chapter were presented in 2003 and 2004 [228, 229]. The only other attempt known to us to calculate the electric field inside a resistive ring carrying a steady current due to charges distributed along the surface of the ring is that due to Weber in 1852 [32]. See the Appendix.

Our goal is to find a solution for the potential due to a current distributed in a finite volume of space, which creates an electric field outside the Ohmic conductor. The only author who has fully solved a problem with the current bounded in a finite volume (beyond the case presented in the previous Chapter) is Jackson [12], who considered a coaxial cable of finite length. But as he considered a return conductor of zero resistivity, he obtained an electric field only inside the cable, with no electric field outside it.

13.2 Description of the Problem

Consider a stationary toroidal Ohmic conductor (greater radius R_0 and smaller radius r_0) with a steady current I, constant over the length $2\pi R_0$ of the conductor. We assume that the conductor has uniform resistivity, and the current is in the azimuthal direction, flowing along the circular loop. The toroid is centered on the plane $z = 0$, z being its axis of symmetry. There is a battery located at $\varphi = \pi$ rad maintaining constant potentials at its extremities. See Fig. 13.1. We initially idealize the battery as of negligible thickness. Later on we consider the battery occupying a finite volume. The medium outside the conductor is supposed to be air or vacuum.

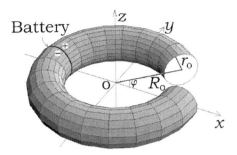

Figure 13.1: A toroidal Ohmic conductor with axis of symmetry along the z axis, smaller radius r_0 and greater radius R_0. A thin battery is located at $\varphi = \pi$ rad maintaining constant potentials (represented by the + and - signs) in its extremities. A steady current flows azimuthally in this circuit loop in the clockwise direction, from $\varphi = \pi$ rad to $\varphi = -\pi$ rad.

The goal here is to find the electic potential ϕ everywhere in space, using the potential at the surface of the conductor as a boundary condition. The problem treated here can be applied to two cases: (a) the toroid is a full homogeneous solid and the battery is a disc. See Fig. 13.2a. And (b) the toroid is hollow and the battery is a circumference. See Fig. 13.2b. The symmetry of this problem suggests the approach of toroidal coordinates (η, χ, φ) see Figure 13.3 [224, p. 112]. These coordinates were introduced by C. Neumann [230], who studied the distribution of surface charges in a metallic ring kept at a constant potential [231, p. 516].

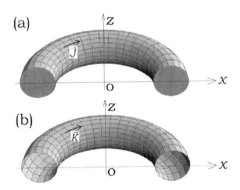

Figure 13.2: The two cases being considered here: (a) a full solid resistive toroidal conductor, with an azimuthal volume current density $\vec{J}$ through the cross-section; (b) a hollow resistive toroidal conductor, with an azimuthal surface current density $\vec{K}$ through the circumference $2\pi r_0$ of the hollow toroidal shell.

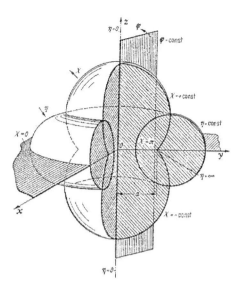

Figure 13.3: Toroidal coordinates (η, χ, φ).

These coordinates are defined by:

$$x \equiv a\frac{\sinh\eta \cos\varphi}{\cosh\eta - \cos\chi}, \qquad y \equiv a\frac{\sinh\eta \sin\varphi}{\cosh\eta - \cos\chi}, \qquad z \equiv a\frac{\sin\chi}{\cosh\eta - \cos\chi}. \tag{13.1}$$

Here a is a constant that gives the radius of a circumference in the $z = 0$ plane described by $\eta \to \infty$ (that is, when $\eta \to \infty$ we have $x = a\cos\varphi$, $y = a\sin\varphi$ and $z = 0$). The values assumed by the toroidal coordinates are: $0 \leq \eta < \infty$, $-\pi$ rad $\leq \chi \leq \pi$ rad and $-\pi$ rad $\leq \varphi \leq \pi$ rad. The inverse transformations are given by:

$$\eta = \operatorname{arctanh}\frac{2a\sqrt{x^2 + y^2}}{x^2 + y^2 + z^2 + a^2}, \qquad \chi = \arctan\frac{2za}{x^2 + y^2 + z^2 - a^2},$$

$$\varphi = \arctan\frac{y}{x}. \tag{13.2}$$

It is also convenient to present here the expressions for $\sinh\eta$, $\cosh\eta$ and $\cos\chi$:

$$\sinh\eta = \frac{2a\sqrt{x^2 + y^2}}{\sqrt{(x^2 + y^2 + z^2 - a^2)^2 + 4a^2z^2}}, \tag{13.3}$$

$$\cosh\eta = \frac{x^2 + y^2 + z^2 + a^2}{\sqrt{(x^2 + y^2 + z^2 - a^2)^2 + 4a^2z^2}}, \tag{13.4}$$

$$\cos\chi = \frac{x^2 + y^2 + z^2 - a^2}{\sqrt{(x^2 + y^2 + z^2 - a^2)^2 + 4a^2z^2}}. \tag{13.5}$$

The surface of the toroid is described by a constant η_0. The internal (external) region of the toroid is characterized by $\eta > \eta_0$ ($\eta < \eta_0$). The greater radius R_0 and the smaller radius r_0 are related to η_0 and to a by $R_0 = a \cosh \eta_0 / \sinh \eta_0$ and $r_0 = a/\sinh \eta_0$. See Figs. 13.1 and 13.3. That is, $R_0/r_0 = \cosh \eta_0$ and $\eta_0 = \cosh^{-1}(R_0/r_0)$.

Laplace's equation for the electric potential ϕ, $\nabla^2 \phi = 0$, has the following form in toroidal coordinates:

$$\nabla^2 \phi = \frac{(\cosh \eta - \cos \chi)^2}{a^2 \sinh \eta} \left[\frac{\partial}{\partial \eta} \left(\frac{\sinh \eta}{\cosh \eta - \cos \chi} \frac{\partial \phi}{\partial \eta} \right) \right.$$
$$\left. + \sinh \eta \frac{\partial}{\partial \chi} \left(\frac{1}{\cosh \eta - \cos \chi} \frac{\partial \phi}{\partial \chi} \right) \right] + \frac{(\cosh \eta - \cos \chi)^2}{a^2 \sinh^2 \eta} \frac{\partial^2 \phi}{\partial \varphi^2} = 0 \,. \quad (13.6)$$

It can be solved in toroidal coordinates with the method of separation of variables (by a procedure known as R-separation), leading to a solution of the form [224, p. 112]:

$$\phi(\eta, \chi, \varphi) = \sqrt{\cosh \eta - \cos \chi} H(\eta) X(\chi) \Phi(\varphi) \,. \quad (13.7)$$

The functions H, X, and Φ which appear here satisfy, respectively, the ordinary equations (with $\Upsilon \equiv \cosh \eta$, and where p and q are constants):

$$(\Upsilon^2 - 1)H'' + 2\Upsilon H' - [(p^2 - 1/4) + q^2/(\Upsilon^2 - 1)]H = 0, \quad (13.8)$$
$$X'' + p^2 X = 0, \quad (13.9)$$
$$\Phi'' + q^2 \Phi = 0. \quad (13.10)$$

13.3 General Solution

The solutions of Eqs. (13.9) and (13.10) for $p \neq 0$ and $q \neq 0$ are linear combinations of the general forms $X_p(\chi) = C_{p\chi} \cos(p\chi) + D_{p\chi} \sin(p\chi)$ and $\Phi_q(\varphi) = C_{q\varphi} \cos(q\varphi) + D_{q\varphi} \sin(q\varphi)$, respectively, where $C_{p\chi}$, $D_{p\chi}$, $C_{q\varphi}$ and $D_{q\varphi}$ are constants. When $p = q = 0$ the solutions reduce to, respectively, $X_0(\chi) = C_{0\chi} + D_{0\chi}\chi$ and $\Phi_0(\varphi) = C_{0\varphi} + D_{0\varphi}\varphi$. Eq. (13.8) is the associated Legendre equation, whose solutions are the associated Legendre functions $P_{p-\frac{1}{2}}^q(\cosh \eta)$ and $Q_{p-\frac{1}{2}}^q(\cosh \eta)$, known as toroidal Legendre polynomials [232, p. 173].

The solution must be periodic in φ, i.e., $\phi(\eta, \chi, \varphi + 2\pi) = \phi(\eta, \chi, \varphi)$, and in χ, i.e., $\phi(\eta, \chi + 2\pi, \varphi) = \phi(\eta, \chi, \varphi)$. This condition implies that $D_{0\varphi} = 0$, $D_{0\chi} = 0$, $q = 1, 2, 3, \ldots$, and $p = 1, 2, 3, \ldots$

The functions $Q_{p-\frac{1}{2}}^q(\cosh \eta)$ are irregular in $\eta = 0$ (which corresponds to the z axis, or to great distances from the toroid). For this reason we eliminate them as physical solutions for this problem in the region outside the toroid (that is, $\eta < \eta_0$). The general solution consists of linear combinations of all possible regular solutions of $P^q_{p-\frac{1}{2}}(\cosh\eta)$, $X_p(\chi)$ and $\Phi_q(\varphi)$:

$$\phi(\eta \leq \eta_0, \chi, \varphi) = \sqrt{\cosh\eta - \cos\chi} \left\{ \sum_{q=0}^{\infty} [C_{q\varphi}\cos(q\varphi) + D_{q\varphi}\sin(q\varphi)] \right.$$

$$\left. \times \left[\sum_{p=0}^{\infty} [C_{p\chi}\cos(p\chi) + D_{p\chi}\sin(p\chi)] P^q_{p-\frac{1}{2}}(\cosh\eta) \right] \right\}. \quad (13.11)$$

We utilized the fact that $\sin 0 = 0$ and $\cos 0 = 1$ to sum up from $p = q = 0$ to ∞. Here $P^0_{p-\frac{1}{2}}(\cosh\eta) \equiv P_{p-\frac{1}{2}}(\cosh\eta)$ are the Legendre functions [202, p. 724].

13.4 Particular Solution for a Steady Azimuthal Current

The surface of the toroid is described by a constant η_0. Here we study the case of a steady current flowing in the azimuthal φ direction along the Ohmic toroid. For this reason we suppose that the potential along the surface of the toroid is linear in φ, $\phi(\eta_0, \chi, \varphi) = \phi_A + \phi_B \varphi / 2\pi$. This potential can be expanded in Fourier series in φ:

$$\phi(\eta_0, \chi, \varphi) = \phi_A + \phi_B \frac{\varphi}{2\pi} = \phi_A + \frac{\phi_B}{\pi} \left[\sum_{q=1}^{\infty} \frac{(-1)^{q-1}}{q} \sin(q\varphi) \right]. \quad (13.12)$$

Fig. 13.4 shows the Fourier expansion of the potential along the conductor surface as a function of φ. The oscillations close to $\varphi = \pm\pi$ rad are due to a Fourier series with a finite number of terms. The overshooting is known as the Gibbs phenomenon, a peculiarity of the Fourier series at a simple discontinuity [202, p. 783–7].

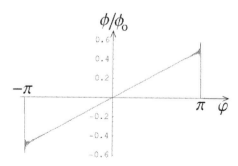

Figure 13.4: Fourier expansion of the potential along the conductor surface as a function of the azimuthal angle φ, Eq. (13.12) with $\phi_A = 0$ and $\phi_B = \phi_0$.

We assume that the potential inside the full solid toroidal Ohmic conductor (that is, for $\eta > \eta_0$), Fig. 13.2a, is also given by Eq. (13.12), namely:

$$\phi(\eta > \eta_0, \chi, \varphi) = \phi_A + \phi_B \frac{\varphi}{2\pi} \,. \tag{13.13}$$

The electric field inside the solid toroid can be expressed in cylindrical coordinates (ρ, φ, z) simply as:

$$\vec{E} = -\nabla \phi = -\frac{\phi_B}{2\pi\rho} \hat{\varphi} \,. \tag{13.14}$$

This electric field does not lead to any accumulation of charges inside a full solid conductor because $\nabla \cdot \vec{E} = 0$.

These are reasonable results. The potential satisfies Laplace's equation $\nabla^2 \phi = 0$, as expected. The electric field is inversely proportional to the distance $\rho = \sqrt{x^2 + y^2}$ from the z axis. This was to be expected as we are assuming a conductor of uniform resistivity. The difference of potential $\Delta\phi$ created by the battery at $\varphi = \pi$ rad can be related to the azimuthal electric field by a line integral:

$$\Delta\phi = -\int_{\varphi=\pi}^{-\pi} \vec{E} \cdot d\vec{\ell} = -E_\varphi 2\pi\rho \,. \tag{13.15}$$

Here ρ is the radius of a circular path centered on the z axis and located inside or along the surface of the toroid. This shows that E_φ should be inversely proportional to ρ, as found in Eq. (13.14). Comparing Eqs. (13.14) and (13.15) yields:

$$\Delta\psi = \phi_B \,. \tag{13.16}$$

By Ohm's law $\vec{J} = g\vec{E}$ (where g is the uniform conductivity of the wire) we can see that $\vec{J}$ is also inversely proportional to the distance ρ from the z axis inside a full solid homogeneous toroidal conductor.

We now consider the solution outside the conductor, valid for the cases of a solid and a hollow toroid.

We calculate Eq. (13.11) with $\eta = \eta_0$ and utilize Eq. (13.12) as a boundary condition of this problem. As we do not have terms with $\cos(q\varphi)$ in Eq. (13.12), this means that $C_{q\varphi} = 0$ for $q = 1, 2, 3, \ldots$ Comparing Eq. (13.11) at $\eta = \eta_0$ with Eq. (13.12) yields two equations connecting ϕ_A and ϕ_B to the C's and D's, namely:

$$\phi_A = C_{0\varphi} \sqrt{\cosh\eta_0 - \cos\chi} \left\{ \sum_{p=0}^{\infty} [C_{p\chi} \cos(p\chi) \right.$$
$$\left. + D_{p\chi} \sin(p\chi)] P_{p-\frac{1}{2}}(\cosh\eta_0) \right\} \,, \tag{13.17}$$

$$\phi_B = \frac{\pi q D_{q\varphi}}{(-1)^{q-1}} \sqrt{\cosh\eta_0 - \cos\chi} \left\{ \sum_{p=0}^{\infty} [C_{p\chi} \cos(p\chi) \right.$$

$$+ D_{p\chi} \sin(p\chi)] P^q_{p-\frac{1}{2}}(\cosh \eta_0) \right\} . \tag{13.18}$$

We now isolate the term $1/\sqrt{\cosh \eta_0 - \cos \chi}$ in Eqs. (13.17) and (13.18), expanding it in Fourier series. That is:

$$\frac{1}{\sqrt{\cosh \eta_0 - \cos \chi}} = \frac{1}{2\pi} \left\{ \sum_{p=0}^{\infty} (2 - \delta_{0p}) \left[\int_{-\pi}^{\pi} \frac{\cos(p\chi')d\chi'}{\sqrt{\cosh \eta_0 - \cos \chi'}} \right] \cos(p\chi) \right\}$$

$$= \frac{\sqrt{2}}{\pi} \left[\sum_{p=0}^{\infty} (2 - \delta_{0p}) Q_{p-\frac{1}{2}}(\cosh \eta_0) \cos(p\chi) \right] , \tag{13.19}$$

where δ_{wp} is the Kronecker delta, which is zero for $w \neq p$ and one for $w = p$. In the last passage we utilized an integral representation of $Q_{p-\frac{1}{2}}(\cosh \eta)$ [232, p. 156, Eq. (10)]:

$$Q_{p-\frac{1}{2}}(\cosh \eta_0) = \frac{1}{2\sqrt{2}} \int_{-\pi}^{\pi} \frac{\cos(p\chi')d\chi'}{\sqrt{\cosh \eta_0 - \cos \chi'}} . \tag{13.20}$$

As in Eq. (13.19) we do not have terms of $\sin(p\chi)$, this means that $D_{p\chi} = 0$ in Eqs. (13.17) and (13.18). Using Eq. (13.19) with Eq. (13.17) yields (for $p = 0, 1, 2, \ldots$):

$$A_p \equiv C_{0\varphi} C_{p\chi} = \frac{\phi_A (2 - \delta_{0p})}{2\pi P_{p-\frac{1}{2}}(\cosh \eta_0)} \int_{-\pi}^{\pi} \frac{\cos(p\chi')d\chi'}{\sqrt{\cosh \eta_0 - \cos \chi'}}$$

$$= \frac{\sqrt{2}\phi_A (2 - \delta_{0p})}{\pi} \frac{Q_{p-\frac{1}{2}}(\cosh \eta_0)}{P_{p-\frac{1}{2}}(\cosh \eta_0)} . \tag{13.21}$$

Using Eq. (13.19) with Eq. (13.18) yields:

$$B_{pq} \equiv D_{q\varphi} C_{p\chi} = \frac{\phi_B (-1)^{q-1}(2 - \delta_{0p})}{2q\pi^2 P^q_{p-\frac{1}{2}}(\cosh \eta_0)} \int_{-\pi}^{\pi} \frac{\cos(p\chi')d\chi'}{\sqrt{\cosh \eta_0 - \cos \chi'}}$$

$$= \frac{\sqrt{2}\phi_B (-1)^{q-1}(2 - \delta_{0p})}{q\pi^2} \frac{Q_{p-\frac{1}{2}}(\cosh \eta_0)}{P^q_{p-\frac{1}{2}}(\cosh \eta_0)} . \tag{13.22}$$

The final solution outside the toroid is given by:

$$\phi(\eta \leq \eta_0, \chi, \varphi) = \sqrt{\cosh \eta - \cos \chi} \left\{ \sum_{p=0}^{\infty} A_p \cos(p\chi) P_{p-\frac{1}{2}}(\cosh \eta) \right.$$

$$+ \sum_{q=1}^{\infty} \sin(q\varphi) \left[\sum_{p=0}^{\infty} B_{pq} \cos(p\chi) P^q_{p-\frac{1}{2}}(\cosh \eta) \right] \right\} , \tag{13.23}$$

where the coefficients A_p and B_{pq} are given by Eqs. (13.21) and (13.22), respectively.

For the region inside the hollow toroid (that is, $\eta > \eta_0$), Fig. 13.2b, we have $P^q_{p-\frac{1}{2}}(\cosh\eta \to \infty) \to \infty$, while $Q^q_{p-\frac{1}{2}}(\cosh\eta \to \infty) \to 0$. For this reason we eliminate $P^q_{p-\frac{1}{2}}(\cosh\eta)$ as physical solutions for the region inside the hollow toroid. The potential is then given by:

$$\phi(\eta > \eta_0, \chi, \varphi) = \phi_A + \sqrt{\cosh\eta - \cos\chi}$$

$$\times \left\{ \sum_{q=1}^{\infty} \sin(q\varphi) \left[\sum_{p=0}^{\infty} B'_{pq} \cos(p\chi) Q^q_{p-\frac{1}{2}}(\cosh\eta) \right] \right\}, \quad (13.24)$$

where the coefficients B'_{pq} are defined by:

$$B'_{pq} \equiv \frac{\phi_B(-1)^{q-1}(2-\delta_{0p})}{2q\pi^2 Q^q_{p-\frac{1}{2}}(\cosh\eta_0)} \int_{-\pi}^{\pi} \frac{\cos(p\chi')d\chi'}{\sqrt{\cosh\eta_0 - \cos\chi'}}$$

$$= \frac{\sqrt{2}\phi_B(-1)^{q-1}(2-\delta_{0p})}{q\pi^2} \frac{Q_{p-\frac{1}{2}}(\cosh\eta_0)}{Q^q_{p-\frac{1}{2}}(\cosh\eta_0)}. \quad (13.25)$$

Note that the potential inside the solid toroid, Eq. (13.13), and the potential inside the hollow toroid, Eq. (13.24), are different. This happens because the discontinuous boundary condition, Eq. (13.12), applies for any $\eta > \eta_0$ inside the solid toroid, particularly for $\varphi \to \pi$ rad ($\phi \to \phi_A + \phi_B/2$) and $\varphi \to -\pi$ rad ($\phi \to \phi_A - \phi_B/2$), where the disc battery is located. See Fig. 13.2a. This does not happen to the hollow toroid, where the battery is a circumference, and the potential must be continuous inside the hollow toroid. See Fig. 13.2b.

We plotted the equipotentials of a full solid toroid on the plane $z = 0$ in Fig. 13.5 with $\phi_A = 0$ and $\phi_B = \phi_0$. We utilized a toroidal surface described by $\eta_0 = 2.187$.

Figure 13.6 shows a plot of the equipotentials of the full solid toroid in the plane $x = 0$ (perpendicular to the current), also with $\phi_A = 0$, $\phi_B = \phi_0$, $R_0 = 1$ and $\eta_0 = 2.187$.

13.5 Potential in Particular Cases

We now analyze the potential outside the toroid, Eq. (13.23), in four regions: (A) far away from the toroid, (B) close to the origin, (C) along the z axis, and (D) along the circumference described by $x^2 + y^2 = a^2$ in the plane $z = 0$.

(A) For great distances from the toroid (that is, $r = \sqrt{x^2 + y^2 + z^2} \gg a$), Eqs. (13.2) to (13.5) yield:

$$\eta \approx \frac{2a\sqrt{x^2+y^2}}{r^2} \ll 1, \quad (13.26)$$

$$\cosh\eta \approx 1 + \frac{2a^2(x^2+y^2)}{r^4} \to 1, \quad (13.27)$$

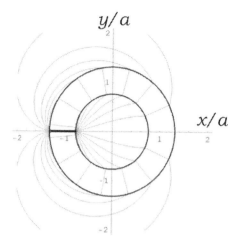

Figure 13.5: Equipotentials for a resistive full solid toroidal conductor in the plane $z = 0$. The bold circumferences represent the borders of the toroid. The current runs in the azimuthal direction, from $\varphi = \pi$ rad to $\varphi = -\pi$ rad. The thin battery is on the left ($\varphi = \pi$ rad). We have used $R_0 = 1$ and $\eta_0 = 2.187$.

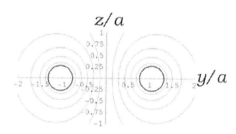

Figure 13.6: Equipotentials in the plane $x = 0$ for a resistive full solid toroidal conductor carrying a steady azimuthal current, Eq. (13.23) with $\phi_A = 0$ and $\phi_B = \phi_0$. The bold circumferences represent the conductor surface. We have used $R_0 = 1$ and $\eta_0 = 2.187$.

$$\cos\chi \approx 1 - \frac{2a^2 z^2}{r^4} \to 1 \;, \tag{13.28}$$

$$\chi \approx \frac{2az}{r^2} \ll 1 \;, \tag{13.29}$$

$$\sqrt{\cosh\eta - \cos\chi} \approx \frac{a\sqrt{2}}{r} \ll 1 \;. \tag{13.30}$$

For $\cosh\eta \approx 1 + \epsilon$, where $0 < \epsilon \ll 1$, we have the following expansion [232, pp. 163 and 173]:

$$P^q_{p-\frac{1}{2}}(1+\epsilon) \approx \frac{\Gamma\left(p+q+\frac{1}{2}\right)}{2^{q/2}q!\,\Gamma\left(p-q+\frac{1}{2}\right)}\epsilon^{q/2}\left\{1+\epsilon\left[\frac{p^2-\frac{1}{4}}{2(1+q)}-\frac{q}{4}\right]\right\}\;. \tag{13.31}$$

That is, for $q=0$ and for $q=1,2,3,\ldots$ we have, respectively:

$$P_{p-\frac{1}{2}}(1+\epsilon) \approx 1 + \epsilon\left(\frac{p^2}{2} - \frac{1}{8}\right) \to 1, \tag{13.32}$$

$$P^q_{p-\frac{1}{2}}(1+\epsilon) \approx \frac{2^{-q/2}\Gamma(p+q+\frac{1}{2})}{q!\Gamma(p-q+\frac{1}{2})}\epsilon^{q/2} \ll 1. \tag{13.33}$$

This means that the terms which appear in the potential for $\eta \ll 1$, up to the order $\epsilon^{1/2}$, are those which have the polynomials with $q=0$ and with $q=1$. That is, $P_{p-\frac{1}{2}}(\cosh\eta \approx 1+\epsilon) \approx 1$ and $P^1_{p-\frac{1}{2}}(\cosh\eta \approx 1+\epsilon) \approx (p^2-1/4)\sqrt{\epsilon/2}$.

The potential ϕ, Eq. (13.23), at great distances from the origin, is given in spherical coordinates (r, θ, φ) by (where $\epsilon = 2a^2\sin^2\theta/r^2$):

$$\phi(r \gg a, \theta, \varphi) \approx \frac{a\sqrt{2}}{r}\left\{\sum_{p=0}^{\infty} \cos\left(p\frac{2a\cos\theta}{r}\right)\right.$$

$$\left.\times \left[A_p + B_{p1}\left(p^2 - \frac{1}{4}\right)\frac{a}{r}\sin\varphi\sin\theta\right]\right\}, \tag{13.34}$$

so that $\phi(r \to \infty) \to 0$, as expected.

(B) The potential close to the origin (that is, $r \ll a$) can be calculated in the same manner. In this approximation:

$$\eta \approx \frac{2\sqrt{x^2+y^2}}{a} \ll 1, \tag{13.35}$$

$$\cosh\eta \approx 1 + \frac{2(x^2+y^2)}{a^2} \to 1, \tag{13.36}$$

$$\cos\chi \approx -1 + \frac{2z^2}{a^2} \to -1, \tag{13.37}$$

$$\chi \approx \pi - \frac{2z}{a} \to \pi, \tag{13.38}$$

$$\sqrt{\cosh\eta - \cos\chi} \approx \sqrt{2} + \frac{x^2+y^2-z^2}{\sqrt{2a^2}} \to \sqrt{2}. \tag{13.39}$$

The potential (13.23) can be expressed as (with $\epsilon = 2r^2\sin^2\theta/a^2$):

$$\phi(r \ll a, \theta, \varphi) \approx \sqrt{2}\left\{\sum_{p=0}^{\infty}(-1)^p \cos\left(p\frac{2r\cos\theta}{a}\right)\right.$$

$$\left.\times \left[A_p + B_{p1}\left(p^2 - \frac{1}{4}\right)\frac{r}{a}\sin\varphi\sin\theta\right]\right\}. \tag{13.40}$$

(C) Along the z axis we have $\sqrt{x^2+y^2} = 0$. From Eqs. (13.2) to (13.5) we have:

$$\eta = 0, \tag{13.41}$$

$$\cosh \eta = 1 \ , \tag{13.42}$$

$$\cos \chi = \frac{z^2 - a^2}{z^2 + a^2} \ , \tag{13.43}$$

$$\sqrt{\cosh \eta - \cos \chi} = a\sqrt{\frac{2}{z^2 + a^2}} \ . \tag{13.44}$$

The potential (13.23) along the z axis can be written as:

$$\phi\left(r = \sqrt{x^2 + y^2 + z^2} = |z|, \theta, \varphi\right)$$

$$= a\sqrt{\frac{2}{z^2 + a^2}} \left[\sum_{p=0}^{\infty} A_p \cos\left(p \arccos \frac{z^2 - a^2}{z^2 + a^2}\right)\right] \ . \tag{13.45}$$

(D) In the circumference described by $x^2 + y^2 = a^2$, along the plane $z = 0$, we have $\eta \to \infty$. The associated Legendre functions $P^q_{p-\frac{1}{2}}(\cosh \eta)$ and $Q^q_{p-\frac{1}{2}}(\cosh \eta)$, for $\eta \gg 1$ (and, therefore, for $\cosh \eta \gg 1$), can be approximated utilizing [232, p. 164]:

$$Q^q_{p-\frac{1}{2}}(\cosh \eta \gg 1) \approx \frac{(-1)^q \sqrt{\pi} \Gamma\left(p + q + \frac{1}{2}\right)}{2^{p+\frac{1}{2}} p! \cosh^{p+\frac{1}{2}} \eta} \ , \quad \text{for any } p \ , \tag{13.46}$$

$$P^q_{p-\frac{1}{2}}(\cosh \eta \gg 1) \approx \frac{2^{p-\frac{1}{2}} (p-1)! \cosh^{p-\frac{1}{2}} \eta}{\sqrt{\pi} \Gamma\left(p - q + \frac{1}{2}\right)} \ , \quad \text{for } p > 0 \ , \tag{13.47}$$

where Γ is the gamma function [202, p. 591]. The potential inside the hollow toroid, Eq. (13.24), assumes the following form along this circumference:

$$\phi(\eta \to \infty, \chi, \varphi) = \phi_A - \frac{\phi_B}{\pi^{3/2}} Q_{-\frac{1}{2}}(\cosh \eta_0) \left[\sum_{q=1}^{\infty} \frac{\sin(q\varphi) \Gamma\left(q + \frac{1}{2}\right)}{q Q^q_{-\frac{1}{2}}(\cosh \eta_0)}\right] \ . \tag{13.48}$$

13.6 Electric Field and Surface Charges

In toroidal coordinates the gradient is written as:

$$\nabla \phi = \frac{1}{a}(\cosh \eta - \cos \chi)\left(\hat{\eta}\frac{\partial \phi}{\partial \eta} + \hat{\chi}\frac{\partial \phi}{\partial \chi} + \frac{\hat{\varphi}}{\sinh \eta}\frac{\partial \phi}{\partial \varphi}\right) \ . \tag{13.49}$$

The electric field can then be calculated by $\vec{E} = -\nabla \phi$, whose components for the region outside the toroid ($\eta < \eta_0$) are given by:

$$E_\eta = -\frac{\sinh \eta \sqrt{\cosh \eta - \cos \chi}}{a} \left\{\sum_{p=0}^{\infty} \cos(p\chi)\left\{A_p \left[\frac{1}{2} P_{p-\frac{1}{2}}(\cosh \eta)\right.\right.\right.$$

$$+ (\cosh\eta - \cos\chi) P_{p-\frac{1}{2}}{}'(\cosh\eta) \Big]$$

$$+ \sum_{q=1}^{\infty} \sin(q\varphi) B_{pq} \left[\frac{1}{2} P_{p-\frac{1}{2}}^q (\cosh\eta) + (\cosh\eta - \cos\chi) P_{p-\frac{1}{2}}^q{}'(\cosh\eta) \right] \Bigg\} \,, \tag{13.50}$$

$$E_\chi = -\frac{\sqrt{\cosh\eta - \cos\chi}}{a} \left\{ \sum_{p=0}^{\infty} \left[\frac{\sin\chi \cos(p\chi)}{2} - p(\cosh\eta - \cos\chi)\sin(p\chi) \right] \right.$$

$$\left. \times \left[A_p P_{p-\frac{1}{2}}(\cosh\eta) + \sum_{q=1}^{\infty} \sin(q\varphi) B_{pq} P_{p-\frac{1}{2}}^q(\cosh\eta) \right] \right\}, \tag{13.51}$$

$$E_\varphi = -\frac{(\cosh\eta - \cos\chi)^{3/2}}{a \sinh\eta} \left\{ \sum_{q=1}^{\infty} q \cos(q\varphi) \left[\sum_{p=0}^{\infty} B_{pq} \cos(p\chi) P_{p-\frac{1}{2}}^q(\cosh\eta) \right] \right\}, \tag{13.52}$$

where $P_{p-\frac{1}{2}}^q{}'(\cosh\eta)$ are the derivatives of the $P_{p-\frac{1}{2}}^q(\cosh\eta)$ relative to $\cosh\eta$.

The electric field inside the full solid toroid ($\eta > \eta_0$) is given simply by:

$$E_\eta = 0\,, \qquad E_\chi = 0\,, \qquad E_\varphi = -\frac{\cosh\eta - \cos\chi}{a \sinh\eta} \frac{\phi_B}{2\pi} = -\frac{\phi_B}{2\pi \sqrt{x^2 + y^2}} \,. \tag{13.53}$$

The total surface charge distribution σ_t that creates the electric field inside (and outside of) the conductor, keeping the current flowing, can be obtained with Gauss's law (by choosing a Gaussian surface involving a small portion of the conductor surface) for the full solid toroid, Fig. 13.2a:

$$\sigma_t(\eta_0, \chi, \varphi) = \varepsilon_0 \left[\vec{E}(\eta < \eta_0) \cdot (-\hat{\eta}) + \vec{E}(\eta > \eta_0) \cdot \hat{\eta} \right]_{\eta_0}$$

$$= \frac{\varepsilon_0 \sinh\eta_0}{a} \left\{ \frac{\phi_A + \phi_B \varphi/2\pi}{2} + (\cosh\eta_0 - \cos\chi)^{3/2} \right.$$

$$\left. \times \left\{ \sum_{p=0}^{\infty} \cos(p\chi) \left[A_p P_{p-\frac{1}{2}}{}'(\cosh\eta_0) + \sum_{q=1}^{\infty} \sin(q\varphi) B_{pq} P_{p-\frac{1}{2}}^q{}'(\cosh\eta_0) \right] \right\} \right\}. \tag{13.54}$$

13.7 Thin Toroid Approximation

Suppose that the toroid is very thin, with its radii described by a greater radius $R_0 = a \cosh\eta_0 / \sinh\eta_0 \approx a$ and smaller radius $r_0 = a/\sinh\eta_0$, such that $r_0 \ll$

R_0. See Fig. 13.1. The surface of the toroid is described by $\eta_0 \gg 1$ and, consequently, $\cosh \eta_0 \gg 1$.

In this approximation, the potential inside the hollow toroid and inside the full solid toroid is given by the same expression, Eq. (13.13). The electric field is given by Eqs. (13.14) and (13.53). This means that there is no distribution of surface charges in the internal surface of a hollow thin toroid.

The Legendre functions of the second kind calculated at $\eta = \eta_0$, given by $Q_{p-\frac{1}{2}}(\cosh \eta_0)$, appear in the coefficients A_p and B_{pq} of the potential outside the toroid, Eqs. (13.21) and (13.22), respectively. As Eq. (13.46), calculated in $\eta = \eta_0$ and for $q = 0$, has a factor of $\cosh^{-p-1/2} \eta_0 \ll 1$, we can neglect all terms in Eq. (13.23) having $p > 0$ compared with the term having $p = 0$. The potential outside the thin toroid ($\eta_0 \gg 1$) can then be written as:

$$\phi(\eta \leq \eta_0, \chi, \varphi) = \sqrt{\frac{\cosh \eta - \cos \chi}{\cosh \eta_0}} \left\{ \phi_A \frac{P_{-\frac{1}{2}}(\cosh \eta)}{P_{-\frac{1}{2}}(\cosh \eta_0)} \right.$$

$$\left. + \frac{\phi_B}{\pi} \left[\sum_{q=1}^{\infty} \frac{(-1)^{q-1}}{q} \sin(q\varphi) \frac{P^q_{-\frac{1}{2}}(\cosh \eta)}{P^q_{-\frac{1}{2}}(\cosh \eta_0)} \right] \right\}. \quad (13.55)$$

It is interesting to find the expressions for the potential and electric field outside but in the vicinity of the conductor (that is, $\eta_0 > \eta \gg 1$). A series expansion of the functions $P^q_{-\frac{1}{2}}(\Upsilon)$ and $P^q_{-\frac{1}{2}}{}'(\Upsilon)$ around $\Upsilon \to \infty$ gives as the most relevant terms [232, p. 173]:

$$P^q_{-\frac{1}{2}}(\Upsilon) \approx \frac{\sqrt{2/\pi}}{\Gamma(1/2-q)} \frac{\ln(2\Upsilon) - \psi(1/2-q) - \gamma}{\sqrt{\Upsilon}}, \quad (13.56)$$

$$P^q_{-\frac{1}{2}}{}'(\Upsilon) \approx \frac{\sqrt{2/\pi}}{\Gamma(1/2-q)} \frac{1}{\Upsilon^{3/2}} \left[1 - \frac{\ln(2\Upsilon) - \psi(1/2-q) - \gamma}{2} \right], \quad (13.57)$$

where $\psi(z) = \Gamma'(z)/\Gamma(z)$ is the digamma function, and $\gamma \approx 0.577216$ is the Euler gamma.

The potential just outside the thin toroid, Eq. (13.55), can then be written in this approximation as (utilizing that $\psi(1/2) + \gamma = -\ln 4$):

$$\phi(\eta_0 \geq \eta \gg 1, \chi, \varphi) = \phi_A \frac{\ln(8 \cosh \eta)}{\ln(8 \cosh \eta_0)}$$

$$+ \frac{\phi_B}{\pi} \left[\sum_{q=1}^{\infty} \frac{(-1)^{q-1}}{q} \sin(q\varphi) \frac{\ln(2 \cosh \eta) - \psi\left(\frac{1}{2} - q\right) - \gamma}{\ln(2 \cosh \eta_0) - \psi\left(\frac{1}{2} - q\right) - \gamma} \right]. \quad (13.58)$$

This equation is valid for $-\pi$ rad $\leq \varphi \leq \pi$ rad, even close to the battery.

The electric field close to the surface of the thin toroid, just outside it, obtained from Eq. (13.58), is given by:

$$E_\eta = -\frac{\sinh \eta}{a} \left\{ \frac{\phi_A}{\ln(8 \cosh \eta_0)} \right.$$
$$\left. + \frac{\phi_B}{\pi} \left[\sum_{q=1}^{\infty} \frac{(-1)^{q-1}}{q} \frac{\sin(q\varphi)}{\ln(2 \cosh \eta_0) - \psi\left(\frac{1}{2} - q\right) - \gamma} \right] \right\}, \quad (13.59)$$

$$E_\chi = 0, \quad (13.60)$$

$$E_\varphi = -\frac{\phi_B}{\pi a} \left[\sum_{q=1}^{\infty} (-1)^{q-1} \cos(q\varphi) \frac{\ln(2 \cosh \eta) - \psi\left(\frac{1}{2} - q\right) - \gamma}{\ln(2 \cosh \eta_0) - \psi\left(\frac{1}{2} - q\right) - \gamma} \right] \quad (13.61)$$

Note that

$$E_\varphi(\eta_0) = -\frac{\phi_B}{\pi a} \left[\sum_{q=1}^{\infty} (-1)^{q-1} \cos(q\varphi) \right] = -\frac{\phi_B}{2\pi a}. \quad (13.62)$$

That is, it coincides exactly with the electric field inside the solid toroid, Eq. (13.14). This is a divergent series presented in Eq. (11.30) which arises from differentiation of a convergent Fourier series, as we discussed in Section 11.4. It can be handled by the average procedure presented in Eq. (11.32).

The surface charge distribution in this thin toroid approximation is given by, from Eq. (13.59):

$$\sigma(\eta_0 \gg 1, \chi, \varphi) = -\varepsilon_0 E_\eta(\eta_0) = \frac{\varepsilon_0 \sinh \eta_0}{a} \left[\frac{\phi_A}{\ln(8 \cosh \eta_0)} \right.$$
$$\left. + \frac{\phi_B}{\pi} \left(\sum_{q=1}^{\infty} \frac{(-1)^{q-1}}{q} \frac{\sin(q\varphi)}{\ln(2 \cosh \eta_0) - \psi\left(\frac{1}{2} - q\right) - \gamma} \right) \right], \quad (13.63)$$

which is also valid for $-\pi$ rad $\leq \varphi \leq \pi$ rad. As there is no surface charge distribution in the internal surface of a hollow thin toroid, this expression means the total surface charge distribution which exists only in the external surface of the (hollow or solid) thin toroid.

In Fig. 13.7 we plotted the density of surface charges σ as a function of the azimuthal angle φ obtained from Eq. (13.63). We can see that σ is linear with φ close to $\varphi = 0$ rad. Close to the battery σ diverges to infinity (that is, $\sigma \to \infty$ when $\varphi \to \pm\pi$ rad). To our knowledge the first to conclude correctly that the surface charge density in a resistive ring carrying a steady current grows toward the battery as a function of the azimuthal angle φ in a pace faster than linearly was Weber in 1852. See the Appendix A.

From Figure 13.7 and Eq. (11.11) we can then write the summation of Eq. (13.63) for a thin toroid and far from the battery (that is, for $\eta_0 \gg 1$ and $\varphi \ll \pi$ rad) as

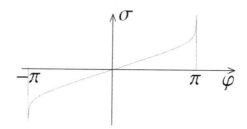

Figure 13.7: Density of surface charges as a function of the azimuthal angle φ obtained from Eq. (13.63) with $\phi_A = 0$ and $\eta_0 = 10$ ($R_0/r_0 = 1.1 \times 10^4$). It is linear with φ when $\varphi \approx 0$ rad but then diverges to infinity close to the battery.

$$\sum_{q=1}^{\infty} \frac{(-1)^{q-1}}{q} \frac{\sin(q\varphi)}{\ln(2\cosh\eta_0) - \psi\left(\frac{1}{2} - q\right) - \gamma} \equiv g(\eta_0)\frac{\varphi}{2} \,. \tag{13.64}$$

Here $g(\eta_0)$ is a dimensionless function of η_0 defined by this equation.

With this definition Eq. (13.63) can be written as

$$\sigma(\eta_0 \gg 1, \chi, \varphi \ll \pi) \approx \frac{\varepsilon_0 \sinh\eta_0}{a} \frac{\phi_A}{\ln(8\cosh\eta_0)} + \frac{\varepsilon_0 \sinh\eta_0}{a} g(\eta_0)\phi_B \frac{\varphi}{2\pi}$$

$$\equiv \sigma_A + \sigma_B \frac{\varphi}{2\pi} \,. \tag{13.65}$$

The constants σ_A and σ_B are defined by this equation, namely

$$\sigma_A \equiv \frac{\varepsilon_0 \sinh\eta_0}{a} \frac{\phi_A}{\ln(8\cosh\eta_0)} \,, \tag{13.66}$$

$$\sigma_B \equiv \frac{\varepsilon_0 \sinh\eta_0}{a} g(\eta_0)\phi_B \,. \tag{13.67}$$

Combining Eq. (13.67) with Eq. (13.62) we can write the tangential component of the electric field E_φ at the surface of the thin toroid as

$$E_\varphi(\eta_0 \gg 1) = -\frac{\phi_B}{2\pi a} = -\frac{\sigma_B}{2\pi\varepsilon_0 \sinh\eta_0 g(\eta_0)} \,. \tag{13.68}$$

As we will see in Appendix A, Weber was the first to obtain an analogous to this result. His approach of dealing with this problem leads to a tangential component of the electric field for a very thin toroid as given by Eq. (A.20), namely:

$$E_\varphi(\eta_0 \gg 1) \approx -\frac{r_0 \sigma_B}{2\pi\varepsilon_0 R_0}\left(\ln\frac{8R_0}{r_0} - \frac{\pi}{2}\right) \,. \tag{13.69}$$

By comparing Eqs. (13.68) and (13.69) for a very thin toroid ($\eta_0 \gg 1$, $a \approx R_0$, $\sinh\eta_0 = a/r_0 \approx R_0/r_0$) we can then try to fit $g(\eta_0)$ as

$$g(\eta_0) \equiv \frac{1}{\ln(R_0/r_0) + K_0} \ . \tag{13.70}$$

The constant K_0 defined by this equation should be a function of η_0 and, according to Eqs. (13.68) and (13.69), should tend to $\ln 8 - \pi/2 = 0.509$ when $\eta_0 \to \infty$.

In Eq. (13.71) we present a least-square fitting of $g(\eta_0)$ given by Eq. (13.64) with 10000 terms in the summation, for φ varying from $-\pi/100$ rad to $\pi/100$ rad, with steps of $\pi/10000$ rad. At the last column we present for each value of $g(\eta_0)$ the corresponding value of K_0 as given by Eq. (13.70).

$$\begin{bmatrix}
\eta_0 & R_0/r_0 & g(\eta_0) & K_0 \\
12.206 & 10^5 & 0.0830051 & 0.534 \\
23.719 & 10^{10} & 0.0424678 & 0.521 \\
35.232 & 10^{15} & 0.0285259 & 0.517 \\
46.745 & 10^{20} & 0.0214746 & 0.515 \\
69.771 & 10^{30} & 0.0143698 & 0.513 \\
92.797 & 10^{40} & 0.0107974 & 0.511 \\
115.822 & 10^{50} & 0.00864749 & 0.511 \\
230.952 & 10^{100} & 0.00433335 & 0.510 \\
461.21 & 10^{200} & 0.00216907 & 0.510
\end{bmatrix} \tag{13.71}$$

This equation indicates that $K_0 \to \ln 8 - \pi/2$, as expected if we apply Weber's approach in order to deal with this problem. See Appendix A. Supposing that this is the case, we can then write the surface charge density for a thin toroid and far from the battery approximately as

$$\sigma(\eta_0 \gg 1, \chi, \varphi \ll \pi) \approx \sigma_A + \sigma_B \frac{\varphi}{2\pi}$$

$$\approx \frac{\varepsilon_0}{r_0} \frac{\phi_A}{\ln(8R_0/r_0)} + \frac{\varepsilon_0}{r_0} \frac{\phi_B}{\ln(R_0/r_0) + \ln 8 - \pi/2} \frac{\varphi}{2\pi} \ . \tag{13.72}$$

In this approximation of a thin toroid, the surface charge density given by Eq. (13.63) does not depend upon the angle χ. This means that the linear charge density $\lambda(\varphi)$ is given simply by $2\pi r_0 \sigma$, namely:

$$\lambda(\eta_0 \gg 1, \varphi) = \frac{2\pi r_0 \varepsilon_0 \sinh \eta_0}{a} \left[\frac{\phi_A}{\ln(8 \cosh \eta_0)} \right.$$

$$\left. + \frac{\phi_B}{\pi} \left(\sum_{q=1}^{\infty} \frac{(-1)^{q-1}}{q} \frac{\sin(q\varphi)}{\ln(2 \cosh \eta_0) - \psi\left(\frac{1}{2} - q\right) - \gamma} \right) \right] \ . \tag{13.73}$$

Far from the battery this reduces to, from Eq. (13.64):

$$\lambda(\eta_0 \gg 1, \varphi \ll \pi) \approx \frac{2\pi r_0 \varepsilon_0 \sinh \eta_0}{a} \frac{\phi_A}{\ln(8 \cosh \eta_0)} + \frac{2\pi r_0 \varepsilon_0 \sinh \eta_0}{a} g(\eta_0) \phi_B \frac{\varphi}{2\pi}$$

$$\equiv \lambda_A + \lambda_B \frac{\varphi}{2\pi} \ . \tag{13.74}$$

The constants λ_A and λ_B were defined by this equation.

We can calculate the total charge q_A of the thin toroid as a function of the constant electric potential ϕ_A. For this end, we integrate the surface charge density σ in χ and φ (in the approximation $\cosh \eta_0 \gg 1$):

$$q_A = \int_{-\pi}^{\pi} h_\chi d\chi \int_{-\pi}^{\pi} h_\varphi d\varphi\, \sigma(\chi,\varphi) = \frac{4\pi^2 \varepsilon_0 \phi_A R_0}{\ln(8\cosh \eta_0)} \approx \frac{4\pi^2 \varepsilon_0 \phi_A R_0}{\ln(8R_0/r_0)} \ , \tag{13.75}$$

where $h_\eta = h_\chi = a/(\cosh \eta - \cos \chi)$ and $h_\varphi = a \sinh \eta/(\cosh \eta - \cos \chi)$ are the scale factors in toroidal coordinates [233]. Notice that from Eq. (13.75) we can obtain the capacitance of the thin toroid [234, p. 127]:

$$C = \frac{q_A}{\phi_A} = \frac{4\pi^2 \varepsilon_0 R_0}{\ln(8\cosh \eta_0)} = \frac{4\pi^2 \varepsilon_0 R_0}{\ln(8R_0/r_0)} \ . \tag{13.76}$$

The potential along the z axis is given by, from Eq. (13.45) in the thin toroid approximation:

$$\phi\left(r = \sqrt{x^2 + y^2 + z^2} = |z|, \theta, \varphi\right) = \frac{q_A}{4\pi\varepsilon_0} \frac{1}{\sqrt{z^2 + a^2}} \ . \tag{13.77}$$

Eq. (13.77) coincides with the coulombian result of a charged thin toroid of radius a in the $z = 0$ plane and total charge q_A.

As we have seen, in the case of a thin toroid the term in the potential with $p = 0$ is much larger than the terms with $p > 0$. This means that Eq. (13.34) reduces to

$$\phi(r \gg a, \theta, \varphi) \approx \frac{a\sqrt{2}}{r} \left[A_0 - \frac{B_{01}}{4} \frac{a}{r} \sin \varphi \sin \theta \right] \ . \tag{13.78}$$

With Eqs. (13.21) and (13.22) we obtain

$$\phi(r \gg a, \theta, \varphi) \approx \frac{a\sqrt{2}}{r} \left[\frac{\sqrt{2}\phi_A}{\pi} \frac{Q_{-\frac{1}{2}}(\cosh \eta_0)}{P_{-\frac{1}{2}}(\cosh \eta_0)} \right.$$

$$\left. - \frac{\sqrt{2}\phi_B}{\pi^2} \frac{Q_{-\frac{1}{2}}^1(\cosh \eta_0)}{P_{-\frac{1}{2}}^1(\cosh \eta_0)} \frac{a}{4r} \sin \varphi \sin \theta \right] \ . \tag{13.79}$$

We now simplify the last two equations utilizing Eqs. (13.46), (13.56) and the relations

$$\Gamma(1/2) = \sqrt{\pi} \ , \quad \Gamma(-1/2) = -2\sqrt{\pi} \ , \tag{13.80}$$

$$\psi(1/2) + \gamma = -\ln 4 \ , \quad \psi(-1/2) + \gamma = 2 - \ln 4 \ . \tag{13.81}$$

This yields:

$$\phi(r \gg a, \theta, \varphi) \approx \frac{a\pi}{r} \left\{ \frac{\phi_A}{\ln(8\cosh\eta_0)} + \frac{\phi_B}{2\pi[\ln(8\cosh\eta_0) - 2]} \frac{a}{r} \sin\varphi \sin\theta \right\}. \tag{13.82}$$

Utilizing a similar procedure beginning with Eq. (13.40) yields:

$$\phi(r \ll a, \theta, \varphi) \approx \pi \left\{ \frac{\phi_A}{\ln(8\cosh\eta_0)} + \frac{\phi_B}{2\pi[\ln(8\cosh\eta_0) - 2]} \frac{r}{a} \sin\varphi \sin\theta \right\}. \tag{13.83}$$

13.8 Comparison of the Thin Toroid Carrying a Steady Current with the Case of a Straight Cylindrical Wire Carrying a Steady Current

It is useful to define a new coordinate system:

$$s' = a\varphi, \qquad \rho' = \sqrt{\left(\sqrt{x^2 + y^2} - a\right)^2 + z^2}. \tag{13.84}$$

We can interpret s' as a distance along the toroid surface in the φ direction, and ρ' as the shortest distance from the circumference $x^2 + y^2 = a^2$ located in the plane $z = 0$. When $\eta_0 > \eta \gg 1$ (that is, $r_0 < \rho' \ll a$), Eqs. (13.84) and (13.4) result in $\cosh\eta \approx a/\rho' \gg 1$ and $\cosh\eta_0 \approx a/r_0 \gg 1$. For $\eta_0 \geq \eta \gg 1$ we can approximate the term inside square brackets of Eq. (13.58) by (taking into account Eq. (11.11)):

$$\sum_{q=1}^{\infty} \frac{(-1)^{q-1}}{q} \sin(q\varphi) \frac{\ln(2\cosh\eta) - \psi\left(\frac{1}{2} - q\right) - \gamma}{\ln(2\cosh\eta_0) - \psi\left(\frac{1}{2} - q\right) - \gamma}$$

$$\approx \left(\sum_{q=1}^{\infty} \frac{(-1)^{q-1}}{q} \sin(q\varphi) \right) \frac{\ln(\cosh\eta)}{\ln(\cosh\eta_0)} = \frac{\varphi}{2} \frac{\ln(\cosh\eta)}{\ln(\cosh\eta_0)}. \tag{13.85}$$

Utilizing Eqs. (13.85) and (13.84) into Eq. (13.58) yields:

$$\phi(\eta_0 \geq \eta \gg 1, \chi, \varphi) = \phi_A \frac{\ln(8a/\rho')}{\ln(8a/r_0)} + \phi_B \frac{s'}{2\pi a} \frac{\ln(a/\rho')}{\ln(a/r_0)}. \tag{13.86}$$

Eq. (13.86) can be written in a slightly different form. Consider a certain piece of the toroid between the angles φ_0 and $-\varphi_0$, with potentials in these extremities given by $\phi_R = \phi_A + \phi_B \varphi_0/2\pi$ and $\phi_L = \phi_A - \phi_B \varphi_0/2\pi$, respectively. This piece has a length of $\ell = 2a\varphi_0$. The potential can then be written as:

$$\phi = \phi_A \frac{\ln(\ell/\rho') - \ln(\ell/8a)}{\ln(\ell/r_0) - \ln(\ell/8a)} + \phi_B \frac{\varphi_0 s'}{\pi \ell} \frac{\ln(\ell/\rho') - \ln(\ell/a)}{\ln(\ell/r_0) - \ln(\ell/a)}$$

$$\approx \left[\frac{\phi_R + \phi_L}{2} + (\phi_R - \phi_L) \frac{s'}{\ell} \right] \frac{\ln(\ell/\rho')}{\ln(\ell/r_0)} \,. \tag{13.87}$$

In the last approximation we neglected the terms $\ln(\ell/8a)$ and $\ln(\ell/a)$ in comparison with the terms $\ln(\ell/\rho')$ and $\ln(\ell/r_0)$ utilizing the approximation $r_0 < \rho' \ll a$ (so that $\ell/r_0 > \ell/\rho' \gg \ell/a$). The electric field can be expressed in this approximation as:

$$\vec{E} = -\left[\frac{\phi_R + \phi_L}{2} + (\phi_R - \phi_L) \frac{s'}{\ell} \right] \frac{\hat{\eta}}{\rho' \ln(\ell/r_0)} - \frac{\phi_R - \phi_L}{\ell} \frac{\ln(\ell/\rho')}{\ln(\ell/r_0)} \hat{\varphi} \,. \tag{13.88}$$

Eqs. (13.87) and (13.88) can be compared to Eqs. (6.17) and (6.18), reproduced as Eqs. (13.89) and (13.90), respectively. These equations refer to a long straight cylindrical conductor of radius r_0 carrying a constant current, in cylindrical coordinates (ρ', φ, z) (note that the conversions from toroidal to cylindrical coordinates in this approximation are $\hat{\eta} \approx -\hat{\rho}'$ and $\hat{\varphi} \approx \hat{z}$). In this case, the cylinder has a length ℓ and radius $r_0 \ll \ell$, with potentials ϕ_L and ϕ_R in the extremities of the conductor, and $RI = \phi_L - \phi_R$:

$$\phi(r \geq a) = \left[\frac{\phi_R + \phi_L}{2} + (\phi_R - \phi_L) \frac{z}{\ell} \right] \frac{\ln(\ell/\rho')}{\ln(\ell/r_0)} \,, \tag{13.89}$$

$$\vec{E}(\rho' \geq a) = \left[\frac{\phi_R + \phi_L}{2} + (\phi_R - \phi_L) \frac{z}{\ell} \right] \frac{\hat{\rho}'}{\rho' \ln(\ell/r_0)} - \frac{\phi_R - \phi_L}{\ell} \frac{\ln(\ell/\rho')}{\ln(\ell/r_0)} \hat{z} \,. \tag{13.90}$$

The potential in the region close to the thin toroid coincides with the cylindrical solution, as expected.

13.9 Charged Toroid without Current

Consider a toroid described by η_0, without current but charged to a constant potential ϕ_A. Using $\phi_B = 0$ in Eqs. (13.23), (13.13) and (13.24) we have the potential inside and outside the toroid, respectively:

$$\phi(\eta \geq \eta_0, \chi, \varphi) = \phi_A \,, \tag{13.91}$$

$$\phi(\eta \leq \eta_0, \chi, \varphi) = \sqrt{\cosh \eta - \cos \chi} \left[\sum_{p=0}^{\infty} A_p \cos(p\chi) P_{p-\frac{1}{2}}(\cosh \eta) \right] \,, \tag{13.92}$$

where $P_{p-\frac{1}{2}}(\cosh \eta_0)$ are the Legendre functions, and the coefficients A_p are given by Eq. (13.21). This solution is already known in the literature [235, p. 239] [236, p. 1304].

It is also possible to obtain the capacitance of the toroid. To this end we compare the electrostatic potential at a distance r far from the origin, Eq. (13.34), with the potential given by a point charge q, $\phi(r \gg a) \approx q/4\pi\varepsilon_0 r$:

$$\phi(r \gg a, \theta, \varphi) \approx \frac{a\sqrt{2}}{r} \left[\sum_{p=0}^{\infty} \frac{\sqrt{2}\phi_A(2 - \delta_{0p})}{\pi} \frac{Q_{p-\frac{1}{2}}(\cosh \eta_0)}{P_{p-\frac{1}{2}}(\cosh \eta_0)} \right] = \frac{q}{4\pi\varepsilon_0 r} . \quad (13.93)$$

The capacitance of the toroid with its surface at a constant potential ϕ_A can be written as $C = q/\phi_A$. From Eq. (13.93) this yields [235, p. 239] [237, p. 5-13] [238, p. 9] [239, p. 375]:

$$C = 8\varepsilon_0 a \left[\sum_{p=0}^{\infty} (2 - \delta_{0p}) \frac{Q_{p-\frac{1}{2}}(\cosh \eta_0)}{P_{p-\frac{1}{2}}(\cosh \eta_0)} \right] . \quad (13.94)$$

Utilizing the thin toroid approximation, $\eta_0 \gg 1$, one can obtain the capacitance of a circular ring, Eq. (13.76).

Another case of interest is that of a charged circular wire already discussed, which is the particular case of a toroid with $r_0 \to 0$. In this case the charged toroid reduces to an uniformly charged circumference of radius $R_0 = a$. With $\eta_0 \gg 1$ and $\cosh \eta_0 \gg 1$ we have $R_0 \approx a$. Keeping only the term with $p = 0$ in Eqs. (13.21) and (13.92) yields (with Eq. (13.75)):

$$\phi(\eta \leq \eta_0, \chi, \varphi) = \phi_A \sqrt{\frac{\cosh \eta - \cos \chi}{\cosh \eta_0}} \frac{P_{-\frac{1}{2}}(\cosh \eta)}{P_{-\frac{1}{2}}(\cosh \eta_0)}$$

$$= \frac{q_A}{4\pi\sqrt{2}\varepsilon_0 a} \sqrt{\cosh \eta - \cos \chi} P_{-\frac{1}{2}}(\cosh \eta) . \quad (13.95)$$

Expressed in spherical coordinates (r, θ, φ), the potential for the thin toroid becomes:

$$\phi(r, \theta, \varphi) = \frac{q_A}{4\pi\varepsilon_0} \frac{1}{[(r^2 - a^2)^2 + 4a^2 r^2 \cos^2 \theta]^{1/4}}$$

$$\times P_{-\frac{1}{2}} \left(\frac{r^2 + a^2}{\sqrt{(r^2 - a^2)^2 + 4a^2 r^2 \cos^2 \theta}} \right) . \quad (13.96)$$

From Eqs. (13.91) and (13.75) we can see that the constant electrostatic potential along the thin toroid expressed in terms of its total charge q_A is given by:

$$\phi(r_0 \ll R_0, \theta, \varphi) = \frac{q_A/2\pi a}{2\pi\varepsilon_0} \ln \frac{8a}{r_0} . \quad (13.97)$$

Even when the linear charge density $q_A/2\pi a$ remains constant, we can see from this expression that the potential diverges logarithmically when $a/r_0 \to \infty$.

We can expand Eq. (13.96) in powers of $r_</r_>$, where $r_<$ ($r_>$) is the lesser (greater) of a and $r = \sqrt{x^2 + y^2 + z^2}$. We present the first three terms:

$$\phi(r,\theta,\varphi) \approx \frac{q_A}{4\pi\varepsilon_0}\left\{\frac{1}{r_>} - \frac{1+3\cos(2\theta)}{8}\frac{r_<^2}{r_>^3}\right.$$

$$\left. + \frac{3}{512}\left[9 + 20\cos(2\theta) + 35\cos(4\theta)\right]\frac{r_<^4}{r_>^5}\right\}. \qquad (13.98)$$

Eqs. (13.95) to (13.98) can be compared with the solution given by Jackson [13, p. 104]. Jackson gives the exact electrostatic solution of the problem of a charged circular wire (that is, a toroid with radius $r_0 = 0$), in spherical coordinates (r, θ, φ):

$$\phi(r,\theta,\varphi) = \frac{q_A}{4\pi\varepsilon_0}\left[\sum_{n=0}^{\infty} \frac{r_<^{2n}}{r_>^{2n+1}} \frac{(-1)^n(2n-1)!!}{2^n n!} P_{2n}(\cos\theta)\right], \qquad (13.99)$$

where q_A is the total charge of the wire. Eq. (13.99) expanded to $n = 2$ yields exactly Eq. (13.98). We have checked that Eqs. (13.96) and (13.99) are the same for at least $n = 30$.

We plotted both Eqs. (13.95) and (13.99), in Fig. 13.8. They yield the same result, as expected. It is worthwhile to note that in spherical coordinates we have an infinite sum, Eq. (13.99), while in toroidal coordinates the solution is given by a single term, Eq. (13.95). The agreement shows that Eqs. (13.95) and (13.99) are the same solution only expressed in different forms.

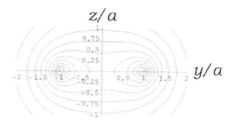

Figure 13.8: Equipotential lines on the plane $x = 0$ (perpendicular to the toroid) for the charged thin wire without current. Both Eqs. (13.95) and (13.99) coincide with one another. We utilized $\eta_0 = 38$ ($\cosh\eta_0 = 1.6 \times 10^{16}$) and $a = 1$. Notice the difference between this Figure and Figure 13.6: the left and right sides of the conductor here possess the same charge signs, while in Figure 13.6 they have opposite signs.

Fig. 13.9 shows the potential as function of ρ (in cylindrical coordinates) in the plane $z = 0$. Eqs. (13.95) and (13.99) give the same result.

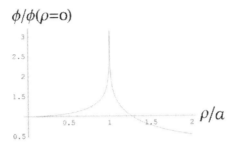

Figure 13.9: Normalized potential as a function of ρ (distance to the z axis) on the plane $z = 0$. Eqs. (13.95) and (13.99) give the same result. We utilized $\eta_0 = 38$ ($\cosh \eta_0 = 1.6 \times 10^{16}$) and $a = 1$.

Along the z axis, i.e., for $\sqrt{x^2 + y^2} = 0$, the potential represented by Eq. (13.96) is given by:

$$\phi(r, \theta, \varphi) = \frac{q_A}{4\pi\varepsilon_0} \frac{1}{\sqrt{z^2 + a^2}} . \qquad (13.100)$$

This is the same result which arises from a direct integration of the electrostatic potential. That is, a charge q_A uniformly distributed along a filiform ring of radius a, located at the plane $z = 0$ and centered along the z axis [202, Example 12.3.3].

13.10 Comparison with Experimental Results

Figure 13.5 can be compared with the experimental result found by Jefimenko [174, Fig. 3], reproduced here in Fig. 13.10 with Fig. 13.5 overlaid on it. The equipotential lines obtained here are orthogonal to the electric field lines. There is a very reasonable agreement between the theoretical result and the experiment.

In order to have a better fit to his data we should consider an extended battery. As we can see from his account of the experiment, Jefimenko painted two sections of his strip with a conducting ink of much smaller resistivity than the remainder of the strip. These sections located at $-\varphi_j < \varphi < -\varphi_i$ and $\varphi_i < \varphi < \varphi_j$ were charged to opposite potentials. Considering these sections as of zero resistivity we can model analytically the potential inside and along the surface of the toroid as:

$$\phi(\eta \geq \eta_0, \chi, \varphi) = \begin{cases} -\phi_B \frac{\varphi_i}{2\pi} \frac{\pi+\varphi}{\pi-\varphi_j}, & -\pi < \varphi < -\varphi_j , \\ -\phi_B \varphi_i/2\pi, & -\varphi_j < \varphi < -\varphi_i , \\ \phi_B \varphi/2\pi, & -\varphi_i < \varphi < \varphi_i , \\ \phi_B \varphi_i/2\pi, & \varphi_i < \varphi < \varphi_j , \\ \phi_B \frac{\varphi_i}{2\pi} \frac{\pi-\varphi}{\pi-\varphi_j}, & \varphi_j < \varphi < \pi . \end{cases} \qquad (13.101)$$

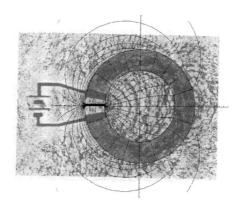

Figure 13.10: Theoretical equipotential lines of Figure 13.5 overlaid on the experimental lines of electric field obtained by Jefimenko. The equipotential lines are orthogonal to the electric field lines.

Notice that the potential described by Eq. (13.101) no longer has a discontinuity at $\varphi = \pi$ rad. The potential is linear between $\varphi = -\varphi_i$ and $\varphi = \varphi_i$, constant for $-\varphi_j < \varphi < -\varphi_i$ and $\varphi_i < \varphi < \varphi_j$, and linear for $-\pi$ rad $< \varphi < -\varphi_j$ and for $\varphi_j < \varphi < \pi$ rad. The boundary condition Eq. (13.12) is now replaced by:

$$\phi(\eta_0, \chi, \varphi) = \frac{\phi_B}{\pi} \left\{ \sum_{q=1}^{\infty} \frac{\sin(q\varphi)}{q^2} \left[\frac{\sin(q\varphi_j)}{\pi - \varphi_j} + \frac{\sin(q\varphi_i)}{\varphi_i} \right] \right\} . \quad (13.102)$$

The potential from Eq. (13.102) is represented in Fig. 13.11 with the values $\varphi_i = 9\pi/10$ rad $= 2.83$ rad and $\varphi_j = 17\pi/18$ rad $= 2.97$ rad. The equipotentials in the plane $z = 0$ are plotted in Fig. 13.12. Fig. 13.13 represents Jefimenko's experiment with Fig. 13.12 overlaid on it. The agreement is now even better than in Fig. 13.10.

Despite this agreement it should be mentioned that Jefimenko's experiment has a conducting strip painted on a glass plate. On the other hand, the theoretical results presented in Figs. 13.5 and 13.12 represent an equatorial slice through a three dimensional toroid. As we saw in Chapter 3, Jefimenko, Barnett and Kelly succeeded in directly measuring the equipotential lines inside and outside a hollow rectangular conductor carrying a steady current. If one day a similar experiment is performed with a toroid, it will be possible to obtain a better comparison with the theoretical results of this Chapter.

The solution inside and along the surface of the full solid toroid yields only an azimuthal electric field, namely, $|E_\varphi| = \Delta\phi/2\pi\rho$. But even for a steady current we must have a component of $\vec{E}$ pointing away from the z axis, E_ρ, due to the curvature of the wire. Here we disregard this component due to its extremely small order of magnitude compared with the azimuthal component E_φ. To grasp this, consider a conducting electron of charge $-e$ and mass m moving azimuthally with drifting velocity v_d in a circumference of radius ρ

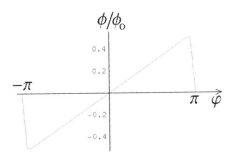

Figure 13.11: Fourier expansion of the potential along the conductor surface as a function of the azimuthal angle φ, Eq. (13.102), with $\phi_B = \pi\phi_0/\varphi_i$. Comparing this Figure with Figure 13.4 we can observe that the oscillations, as well as the overshooting, do not appear anymore, as the potential is now continuous for 0 rad $\leq \varphi \leq 2\pi$ rad. We have used $\varphi_i = 9\pi/10$ rad $= 2.83$ rad and $\varphi_j = 17\pi/18$ rad $= 2.97$ rad.

around the z axis. In a steady state situation there will be a redistribution of charges along the cross-section of the toroid creating an electric field E_ρ which will exert a centripetal force on the conduction electrons. By Newton's second law of motion we can equate the force eE_ρ with the mass of the electron times its centripetal acceleration, in such a way that $eE_\rho = mv_d^2/\rho$. Suppose we have a 14 gauge copper wire ($r_0 = 8.14 \times 10^{-4}$ m) of 1 m length bent in a circumference of radius $R_0 = \rho = (1/2\pi)$ m $= 1.59 \times 10^{-1}$ m carrying a current of 1 A. The drifting velocity is given by $v_d = 3.55 \times 10^{-5}$ m/s, the resistance of the wire is 8.13×10^{-3} Ω and the potential difference created by the battery is $\Delta\phi = 8.13 \times 10^{-3}$ V. This yields $E_\varphi = 8.13 \times 10^{-3}$ V/m and $E_\rho = 4.5 \times 10^{-20}$ V/m. That is $E_\rho \ll E_\varphi$, which justifies disregarding the E_ρ component of the electric field in comparison with the E_φ component.

As we saw in Section 6.4, a stationary conductor carrying a steady current which is uniform over its cross-section generates a charge distribution inside the conductor. This charge distribution creates a radial electric field inside the conductor. In steady state there is then an electric force acting upon any specific conduction electron which is counteracted by the radial magnetic force that arises due to the movement of the other conduction electrons, the radial Hall effect. However, this electric field is rather small, (10^{-5} smaller than the electric field that maintains the current flowing, supposing a typical copper conductor with 1 mm diameter and 4×10^{-3} m/s drifting velocity). For this reason this electric field and the corresponding charge redistribution have been neglected in these calculations.

In this Chapter we presented a solution for the potential inside and outside a resistive toroidal conductor carrying a steady azimuthal current. The current flows in a finite volume of space and the solution obtained here indicates the existence of the electric field outside the conductor. The theoretical calculations were compared to the experimental results, indicating a very good agreement.

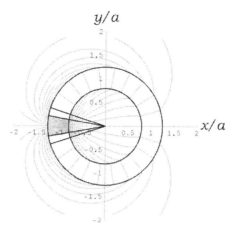

Figure 13.12: Equipotentials in the plane $z = 0$ for a resistive toroidal conductor carrying a steady azimuthal current, using Eq. (13.102) as boundary condition and $\phi_B = \pi\phi_0/\varphi_i$. The bold circumferences represent the conductor surface and the bold straight lines represent the angles $\varphi = \pm\varphi_i = \pm 9\pi/10$ rad = 2.83 rad and $\varphi = \pm\varphi_j = \pm 17\pi/18$ rad = 2.97 rad. We have used $\eta_0 = 2.187$.

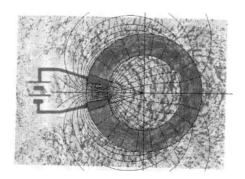

Figure 13.13: Jefimenko's experiment with Figure 13.12 overlaid on it – the equipotential lines are orthogonal to the electric field lines.

We also obtained the distribution of charges along the surface of the resistive ring carrying a steady current, a subject which was first considered by Wilhelm Weber 150 years ago, as we see in the Appendix. Weber made the first preliminary quantitative calculations related to this problem, and this specific case has essentially been forgotten these many years. This Chapter can be seen as a fulfillment of one of Weber's goals. That is, to derive the distribution of surface charges in a ring which, together with the battery, creates a constant tangential electric field for all azimuthal angles inside the ring. We have also succeeded in deriving the force exerted upon a stationary and external point charge by this stationary ring carrying a steady current.

Part IV

Open Questions

Chapter 14

Future Prospects

In this book we presented the main simple cases which can be treated analytically. The goal now might be to consider theoretically other situations which have already been analyzed experimentally. Examples include the current flowing in a disc, Figure 3.7; current-carrying wedges with the two halves connected in parallel and in series, Figure 3.6; *etc.* The latter situation is interesting in order to know quantitatively the correct distribution of charges allowing the current to bend around a corner. Studies along these lines include Rosser [167], Jefimenko [240] [176, pp. 302-303] and the book by Chabay and Sherwood [165, Chapter 6].

Other interesting aspects are connected with the distribution of surface charges close to the battery and inside it. In Chapters 11 to 13 we discussed this in the cases of a cylindrical shell, a spherical shell and a ring with azimuthal currents. Another important discussion for the case of a coaxial cable of finite size has been given by Jackson [12]. Saslow considered a spherical battery surrounded by a conducting medium and analyzed the distribution of charges upon the surface of the battery [241]. The distribution of surface charges close to a battery in the case of straight conductors carrying steady currents has been treated for two different configurations in 2004 [205] and 2005 [222]. Other cases should also be studied quantitatively in different configurations. Although it might be difficult to obtain detailed information analytically about the distribution of surface charges in a battery of finite size, this might be accomplished with computer calculations and numerical plots.

It would also be important to analyze cases in which the current is not generated by a chemical battery, but by the relative motion between a closed conducting circuit and a magnet, as in the first case considered qualitatively by Weber and described in the Appendix. Calculations have been performed relative to the surface charges in the case of a square circuit in the presence of a variable magnetic flux [242], and also the case of a ring rotating in the presence of a magnetic field [243]. It would be important to extend the calculations to other spatial configurations and analogous situations.

Another situation which has received little attention up to now is the distribution of charges in resistive conductors carrying steady currents when these conductors are composed of two or more different materials. That is, the charges that accumulate on the interface between a conductor and a resistor, or on the interface of two conductors with different resistivities. Some authors who have considered this problem include Jefimenko [240], Heald [226], Härtel [244, 245], Chabay and Sherwood [165, 166], and Jackson [12]. Jackson's work has an interesting comparison of the distribution of surface charges in a circuit with a large resistance and in an equivalent open circuit, *i.e.*, with the resistor removed from the circuit.

Another relevant topic is to consider in detail the behavior of surface charges and the corresponding external electric field in the transition from steady-currents to low and high frequency circuits with alternating currents. Important discussions of this subject have been given by Jackson [12] and Preyer [246]. Weber and Kirchhoff's works related with the telegraphy equation discussed in the Appendices should also be reconsidered and extended to different cases and configurations [30, 31].

Beyond these future extensions, there are a number of topics which still need to be clarified. Consider a stationary point charge close to a stationary permanent magnet. Is there a net force between them beyond the force due to electrostatic origin? That is, is there a force depending upon the magnetization of the magnet, or depending upon the magnetic field it produces? As we have seen in this book, there is a force between a stationary point charge and a stationary resistive circuit carrying a steady current. This force is proportional to the electromotive force of the battery. Is there a similar force between a stationary magnet and a stationary external charge? In this question we are not including the force due to electrostatic induction which must exist between a conducting magnet and the external charge, which is of electrostatic origin (due to image charges, *etc.*) The magnet we are considering here has permanent magnetization. Its magnetic field is due to permanent microscopic or molecular currents in its interior. The magnet is not connected to a chemical battery and for this reason it should not have a distribution of surface charges as in the case of a resistive wire carrying a steady current. In any event this subject should be better analyzed and careful experiments should be performed to answer this question.

Analogously, there should not exist an electric field outside a wire made of a superconducting material if it carries a steady current without any external source of electromotive force, *i.e.*, if there is no battery connected to the wire. As this wire has no resistance and is not connected to any battery, there should be no electric field outside the wire (except for the zeroth order electric field if we approach a test charge to the wire). But it should be emphasized once more that only experiments can decide this question.

A possible connection between the external electric field around a resistive cylindrical conductor carrying a steady current and the Aharonov-Bohm effect was discussed in 2001 [247]. Although this idealized infinite conductor will not produce any external magnetic field, it will produce an external electric field if the solenoid is connected to a chemical battery. This electric field is not considered by most authors, as they are unaware of its existence. For this reason none of them considered the influence of this steady electric field in the Aharonov-Bohm effect, taking into account only the magnetic vector potential. The goal of our paper was to call attention to this external electric field for the analysis of the Aharonov-Bohm effect.

In this book we have shown that a force must exist between a point charge and a resistive wire carrying a steady current when they are at rest relative to one another. It has been shown theoretically that this force (or the electric field outside the wire) is proportional to the emf of the battery. But it is still necessary to show experimentally the proportionality between this force and the voltage of the battery. This proportionality should appear according to the calculations presented here. They have yielded qualitative agreement with the experiments of Bergmann, Schaefer, Jefimenko, Barnett and Kelly relating to equipotentials and electric field lines. But we are not not aware of any experiment showing directly the proportionality between this force and the voltage generated by the battery.

Another crucial question which still needs to be settled empirically is related to the second order electric field (proportional to the square of the current, or to the square of the drifting velocity of the mobile electrons). Alternatively we might ask if there is a second order force between a stationary charge and a stationary wire carrying a steady current. This electric field and the corresponding force produced by it upon stationary charges are usually much smaller than the electric field and forces discussed in this book (proportional to the voltage of the battery). For this reason it is difficult to decide unambiguously whether this effect exists. Experiments to decide this question should be performed separately, considering three cases: (1) resistive wires connected to batteries and carrying steady currents, (2) superconductors carrying steady currents without any external source of electromotive force, and (3) permanent magnets. It may happen that this second order electric field exists (or does not exist) for all three cases. It may also be that it exists for one or more of these cases, but not for the other case(s). These three cases must be considered independently from one another. The theoretical analysis of the experiments must take into account the force due to electrostatic induction (zeroth order electric field) and also the component of the electric field discussed in this book proportional to the emf of the battery (for the case of resistive conductors). This is not a simple task in complicated configurations. It is essential to be extremely careful with all possible influences in order to avoid misleading conclusions.

Another topic which has not been treated in this book is the convenience and importance of the surface charges and of microscopic aspects of current conduction for the understanding of the macroscopic phenomena associated with circuits carrying steady currents. This subject has great conceptual and didactic relevance. Several studies relating to the teaching of electromagnetism have been developed through an exploration of this topic, as applied to high school and to university courses [164, 244, 245, 248, 211, 165, 171, 249, 250, 166].

This book has shown how a very simple question of basic electromagnetism has been answered incorrectly by many important authors along several decades.

This has had a negative influence on the development of the subject for more than a century, and created many prejudices which are very difficult to eliminate. We should try to avoid the same mistake in the future. This was one of our reasons for writing this book.

Another goal was to obtain the densities of charges spread upon the surfaces of resistive conductors carrying steady currents in several configurations. For long, straight conductors it was shown that these surface densities are a linear function of the longitudinal coordinate. For curved conductors, on the other hand, they grow faster than linearly along the length of the conductor, increasing their magnitude toward both extremities of the battery. All of this can indeed be understood in terms of electrodynamical principles. Following French in the last page of his didactic book *Newtonian Mechanics* [251, p. 700], the best way to close this work is with a simple and fair statement, namely: "But Weber got there first!"

Appendix A

Wilhelm Weber and Surface Charges

Wilhelm Eduard Weber (1804-1891) was one of the first to mention and analyze quantitatively the surface charges in resistive conductors carrying steady currents. Here we discuss some parts of his papers dealing with this topic. In Section 1.4 we presented some important aspects of his life and work, quoting the publication of his collected papers and all of his works which have been translated into English.

Weber wrote eight major memoirs between 1846 and 1878 under the general title *Electrodynamic Measurements*, or *Determination of Electrodynamic Measures* (the eighth memoir was published only posthumously in his collected papers).

The work which we discuss here is the second memoir of this series, published in 1852: Electrodynamic Measurements Relating Specially to Resistance Measurements [32]. To the best of our knowledge this work has never been translated into English or any other language. What we quote here is our translation. The paper is divided into six parts and has five extra appendices. What interests us here is the fifth part, which extends from Section 28 to Section 36 (pp. 368 to 405 of Vol. 3 of Weber's *Werke* [38]): *On the Connection of the Theory of the Galvanic Circuit with the Electrical Fundamental Laws*. Between square brackets we offer our interpretation of expressions or sentences from Weber. We have produced the figures presented in this Appendix in order to illustrate Weber's reasoning. The footnotes presented here are also ours.

Section 28 begins with the statement that until then there was no development of the relation between the theory of the galvanic current [Ohm's law] and the electrical fundamental laws [Coulomb's force], as these two subjects were treated independently from one another. He says that the main reason for this separate treatment lies in the mathematical difficulty of connecting the two subjects in a complete manner.[1] His goal is to discuss some aspects which can

---
[1] An example of this mathematical difficulty can be seen in Chapter 13 of this book, a

lead to a connection between both subjects. He mentions Ohm's law, valid for steady currents, relating the current intensity, the resistance and the electromotive force (or electro-motor force) [due to a chemical battery, for instance].[2] He remarks that Ohm tried to base his law on the variable volume density of charges in the conductor, in analogy with Fourier's treatment of the propagation of heat based on the variable distribution of the temperature inside a body. That is, in a region of the wire carrying a steady current where there is no electromotive force (no point of contact between two different metals, for instance), the force moving the charges against resistance would be due to a gradient in the volume density of charges.[3] Weber states that Ohm found the key to explain the law of the galvanic circuit based upon the distribution of electric charges in the conductor. On the other hand, he mentions that Ohm's approach is in contradiction with the fundamental laws of electrostatics, according to which free electricity can exist only along the *surface* of a conductor. Weber mentions that the same must be true in the case of a galvanic circuit with steady current, even disregarding the relative motion between the interacting charges. While the local temperature gradient is a necessary condition for the local propagation of heat, the same does not need to be true for charges, as they act at a distance.

Weber then considers an interesting example of a stationary homogeneous closed copper ring with overall equal cross-section. He imagines a magnet moving along the axis of the ring, perpendicular to its plane. See Figure A.1. Weber had already considered briefly this situation in his first major Memoir of 1846, [137, p. 203 of the *Werke*]. According to Weber the magnet will exert the same electromotive force in all elements of the ring. As all elements have the same resistance, the electromotive force will produce the same current in all of them. In this case there will not appear any accumulation of charges in any place of the ring.[4] According to Weber, only when there is a difference of the action of the electromotive force in different parts of the circuit there will appear accumulation of charges.[5] The effect of the distribution of free electricity along the surface of the wire will be to equalize [in all parts of the wire] the action of the electromotive force [due to the contact of two different metals or due to

---

situation which Weber also considered quantitatively in his memoir, as we will see.

[2]In some examples it is possible to understand Weber's *elektromotorische Kraft* as potential difference or as electromotive force (emf) around a complete circuit carrying a steady current. On the other hand, in other situations it seems that Weber refers to the longitudinal component of the electric field driving the conduction charges along a wire carrying a steady current. The electric force associated with this electric field acts in the opposite direction of the resistive frictional force exerted upon the conduction electrons by the crystaline lattice of the metal.

[3]This can be seen in pp. 402 and 418 of Ohm's work [252].

[4]For an experimental proof of this fact, see the interesting paper of Moreau and collaborators [184]. The relative motion between the magnet and the circuit drives the current around the ring due to a non Coulomb force. In this case there is no potential difference between any two points on the ring [204].

[5]This will be the case, for instance, when there is steady current in a resistive wire connected to a chemical battery. The electromotive force of the battery acts mainly inside itself and in the region close to its surface, so that along the other parts of the wire there must be forces of another origin moving the mobile charges against resistive forces.

a chemical battery, this electromotive force having differing intensities in different portions of the circuit]. Weber then states that two things remain to be shown: (1) how this distribution of free electricity is possible according to the fundamental electrical laws, and what its properties should be, [6] and (2) how the surface charges arise and are maintained.

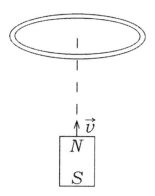

Figure A.1: Magnet moving along the axis of a copper ring. According to Weber the magnet will exert the same electromotive force in all elements of the ring.

Section 29 is entitled "Proof of the possibility of a distribution of the free electricity in a conductor, through which is balanced the difference in the action of given electromotive forces in different parts of the circuit according to the proportionality of their resistances." He begins by considering particles of free electricity along the surface of a conductor exerting electromotive forces [in this case electrostatic forces due to Coulomb's law] upon all charged particles of the conductor. These forces due to surface charges will decrease or increase the electromotive forces of the circuit [due to a chemical battery, for instance]. He then asks if a distribution of surface charges is possible such that the [net] electromotive forces [that is, the resultant electric field due to the chemical battery and to the surface charges] will be equilibrated in all parts of the circuit in proportion with the resistance of these parts. Disregarding the effect of the relative motion between the charges [relative motion between the conduction charges and the ions of the lattice], Weber mentions that this question must be answered based upon the fundamental law of electrostatics. He then mentions the theorem proved by Poisson that there is one and only one possible distribution of charges on the surface of a conductor which equilibrates the electric forces exerted by external charges.

He then applies this theorem conceptually to a cylindrical conductor acted upon by an external point charge along its axis, at a great distance from the cylinder. See Figure A.2. This external charge exerts essentially the same axial electrostatic force on all charges of the cylinder. There will be a redistribution

---

[6]That is, how is it possible to derive from Coulomb's force the distribution of the surface charges which will equalize the electric field in all points inside the resistive wire.

of charges along the surface of the cylinder, creating an opposite electric field and canceling this external force at all internal points of the cylinder. If we now consider the presence of this fixed distribution of surface charges, without the presence of the external point charge [that is, as if the surface charges had been glued upon the surface of the cylinder and later on the external point charge were removed], there will be an axial electromotive force acting on all points of the cylinder.[7]

Next he considers a curved cylinder [like a piece of a ring in the form of an arc of a circle] and a point charge at a great distance from it, along the tangent to one element of the arc. This external point charge exerts a uniform force along the tangent of the arc, which is equilibrated by the force due to the distribution of surface charges in the curved cylinder. When this distribution of surface charges is kept fixed at their places [by the application of other external forces to them] and the external point charge is removed, only the longitudinal electromotive force [acting upon all charges of the cylinder] due to the surface charges will remain. He then generalizes this to all elements of the curved cylinder such that the surface charges on any specific element will be a function of the surface charges on all other elements of the arc.

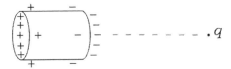

Figure A.2: Point charge q along the axis of a finite cylindrical conductor, at a great distance from it. This external charge exerts essentially the same electrostatic axial force upon all charges of the cylinder. In equilibrium there will be an electrical polarization of the cylinder, with the charges along its surface canceling exactly, at all internal points of the cylinder, the electric field due to the external charge.

He then imagines this curved cylinder making a circle, like a ring, with a small separation between the initial and final cross-sections of the ring. Weber shows that these two surfaces should not touch one another; otherwise there will be an infinite amount of opposite charges on these surfaces (supposing a uniform tangential electromotive force acting on all points of the ring). He calls δ the distance between the extremities of the open ring and $\pm e$ the charges of two elements of these opposite faces. Utilizing Coulomb's law he shows that the force on a test charge inside the ring due to the two opposite faces is proportional to δe. As he wants this electromotive force [or electric field, as we would say today] to remain constant as $\delta \to 0$, it is necessary that simultaneously $e \to \infty$, which was what he wanted to prove. The electromotive force along the ring is

---

[7]He has proved in this first simple case that there is a distribution of surface charges which exerts an equal longitudinal force on all points of a cylinder, although he did not explicitly attempt to calculate the distribution of surface charges in this specific example.

then also proportional to δe. In the open region between the two extremities the electromotive force due to the charges in the end surfaces points in a direction opposite to the direction of the electromotive force acting on a test charge inside the ring and close to the extremities. He concludes that if we want the same electromotive force [net electric field] at all points of the closed ring, then in the region between the extremities it is necessary for an electromotive force to act, independent of the distribution of surface charges [that is, a force of non-electrostatic origin]. As an example of such a force he mentions the case of copper and zinc touching one another [we might also mention the case of a chemical battery].

He draws three conclusions from these considerations:

1. It is not possible to have [steady] current in a closed ring due only to a distribution of surface charges on the ring. It is necessary to have an electromotive force of different origin in at least one cross-section of the ring (like the contact of copper and zinc).[8]

2. The current in a circuit is proportional to the density of surface charges along the circuit.[9] The electromotive force is proportional to δe and to the current in the circuit.

3. When we double all dimensions of a circuit but keep the same electromotive force, then the density of surface charges should remain constant, even though the surface area is four times the previous one.[10] At the same time it follows that when we double all dimensions of a circuit, the distance δ should also double, but when the charge e remains constant, the electromotive force proportional to δe should also double. This double electromotive force requires the same motion [velocity] of the charges in a circuit with doubled dimensions, as [the velocity] in a circuit of simple length and cross-section. But this same motion [velocity] generates four times the current in a circuit with doubled dimensions (and four times the [area of the] cross-section). That is, a doubled electromotive force generates, in a circuit of doubled length and four times the cross-section [in comparison with the simple original circuit], a current four times larger, which is in agreement with the laws of the galvanic circuit.

Section 30 is entitled "On the law of the distribution of the free electricity

---

[8]This is similar to the theorem that $\oint \vec{E} \cdot d\vec{\ell} = 0$, where $\vec{E}$ is the electrostatic field of Coulomb's law, $d\vec{\ell}$ is an element of length and the line integral is over a closed circuit of arbitrary form. That is, in order to have an electromotive force driving a current around a closed resistive circuit it is necessary to have a source of non-electrostatic origin. See Section 5.1.

[9]An example of this general conclusion can be seen in Eq. (6.17) for the case of a straight wire. Combining it with Eq. (6.2), $\sigma(z) = \sigma_A + \sigma_B z/\ell$, yields: $I = -(a\sigma_B/R\varepsilon_0)\ln(\ell/a)$. That is, I is directly proportional to σ_B, as Weber concluded.

[10]From Eqs. (6.2), (6.14) and with the electric field (Weber's electromotive force in this case) given by $E_1 = \Delta\phi/\ell = RI/\ell$ we obtain:

$$\sigma_B = \frac{\varepsilon_0 E_1}{\ln(\ell/a)} \frac{\ell}{a} . \tag{A.1}$$

That is, σ_B is proportional to E_1 and to $(\ell/a)/\ln(\ell/a)$. If at the same time we double ℓ and a, keeping a constant E_1, then σ_B will remain constant. This is an example of Weber's conclusion.

over the surface of a conductor carrying a constant and uniform current." For a linear conductor he says that we can consider the surface charges as distributed along its axis.[11] He shows this considering a cylindrical conductor of length 2λ with a circular cross-section of radius $\alpha \ll \lambda$. See Figure A.3.

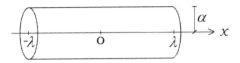

Figure A.3: Cylindrical conductor of length 2λ and radius $\alpha \ll \lambda$. In the case of steady currents, the surface charge density is linear with the longitudinal x component, i.e., proportional to $a + bx$.

He considers initially that in the case of a steady current the density of surface charges is linear with the longitudinal x component, i.e., proportional to $a + bx$.[12] He integrates the longitudinal electromotive force [our electric field along the direction of the axis] due to these surface charges acting on a point located at the origin (the center of the cylinder), obtaining the result (supposing $\lambda \gg \alpha$):

$$\int_{x=-\lambda}^{\lambda} \frac{2\pi\alpha(a+bx)x\,dx}{(\alpha^2 + x^2)^{3/2}} \approx 4\pi\alpha b \left(\log \lambda - \log \frac{e\alpha}{2} \right), \quad (A.2)$$

where $e = 2.7183$ is the natural logarithm base.[13]

He then shows that the same result is obtained when we consider all the surface charges distributed along the axis of the cylinder, integrating from $x = -\lambda$ to $x = \lambda$, with the exception of the region between $x = -e\alpha/2$ and $x = e\alpha/2$. See Figure A.4.

That is, he was able to derive Eq. (A.2) by assuming all surface charges concentrated along the axis of the wire and calculating the longitudinal electric field at the origin integrating from $x = -\lambda$ to $x = -e\alpha/2$ and from $x = e\alpha/2$ to $x = \lambda$.[14]

---

[11] That is, the actual force exerted by the free charges distributed along the surface of a cylindrical conductor carrying a steady current upon a test charge can be replaced by the force upon this test charge due to an appropriate distribution of charges along the axis of the cylinder.

[12] Weber's 2λ and α are equivalent, respectively, to our ℓ and a of Figure 6.1. Weber's $a + bx$ is equivalent to our $\sigma(z) = \sigma_A + \sigma_B z/\ell$, Eq. (6.2).

[13] This result is equivalent to Eq. (6.12), namely, $E_1 = (a\sigma_B/\ell\varepsilon_0) \ln(\ell/ea)$. All results obtained by Weber in this Section can be put in the international system of units by dividing them by $4\pi\varepsilon_0$. Weber's log has base e, which means that his log can be written as our ln. In Chapter 6 we first calculated the potential by integration, and then the electric field by $\vec{E} = -\nabla\phi$. Here Weber has integrated the electric field directly. The final result was the same, as expected.

[14] In other words, he considers the line having a linear charge density given by $2\pi\alpha(a+bx)$. He considered the test charge at the origin. This is a very interesting technique which greatly simplifies the integrations. We have checked his integration and it is correct.

We now generalize his calculation to obtain the longitudinal electric field at an arbitrary

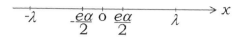

Figure A.4: Weber considered now all surface charges distributed along the axis of the cylinder of length 2λ. According to Weber, the linear integration that yields the same electric field at the origin as that given by Figure A.3 and Eq. (A.2), now runs from $x = -\lambda$ to $x = \lambda$, except in the region between $x = -e\alpha/2$ and $x = e\alpha/2$.

point x', $E_{x'}(x')$:

$$E_{x'}(x') = \left(\int_{x=-\lambda}^{x'-e\alpha/2} + \int_{x=x'+e\alpha/2}^{\lambda} \right) \frac{2\pi\alpha(a+bx)(x'-x)dx}{[(x'-x)^2]^{3/2}}$$

$$= -2\pi\alpha \left[b \ln \frac{4(\lambda^2 - x'^2)}{e^2\alpha^2} - \frac{2(a+bx')x'}{\lambda^2 - x'^2} \right]. \quad (A.3)$$

At $x' = 0$ this yields Weber's result, namely

$$E_{x'}(0) = -4\pi\alpha b \ln \frac{2\lambda}{e\alpha}. \quad (A.4)$$

We now present an alternative way of obtaining the electric field. This alternative procedure will be followed by Weber in the calculation of the ring, as we will see shortly.

If he had wished to obtain the electric field from the potential, he would have had to calculate the potential at a generic point x' (and not only at the origin $x' = 0$). The integrals would need to go from $x = -\lambda$ to $x = x' - e\alpha/2$ and from $x = x' + e\alpha/2$ to $x = \lambda$. Let us write as $\Phi(x')$ the function which would represent the potential at x' calculated in this way (later on we show that it is different from the real potential $\phi(x')$). With a charge element $dq = 2\pi\alpha(a+bx)dx$ we would obtain:

$$\Phi(x') = \left(\int_{x=-\lambda}^{x'-e\alpha/2} + \int_{x=x'+e\alpha/2}^{\lambda} \right) \frac{2\pi\alpha(a+bx)dx}{\sqrt{(x'-x)^2}}$$

$$= 2\pi\alpha \left[(a+bx') \ln \frac{4(\lambda^2 - x'^2)}{e^2\alpha^2} - 2bx' \right]. \quad (A.5)$$

From this expression we obtain

$$-\frac{\partial \Phi(x')}{\partial x'} = -2\pi\alpha \left[b \ln \frac{4(\lambda^2 - x'^2)}{e^2\alpha^2} - \frac{2(a+bx')x'}{\lambda^2 - x'^2} - 2b \right]. \quad (A.6)$$

And this is different from the electric field given by Eq. (A.3)! For instance, in the limit when $x' \to 0$ Eq. (A.6) yields $-4\pi\alpha b \ln(2\lambda/e^2\alpha)$. And there is a difference of $1/e$ inside the logarithm as compared with the previous result which Weber obtained by direct integration of the electric field.

That is, although in general $\vec{E} = -\nabla\phi$, in this particular case we did not obtain $E_{x'}(x') = -\partial\Phi(x')/\partial x'$, as might be expected. The origin of this difference is not easy to locate but we need to clarify it before proceeding. Everything is due to Weber's peculiar approximation method when we calculate $E_{x'}(x')$ or $\Phi(x')$. In this method the location of the point of observation, x', appears not only in the integrand, but also in the limits of the integrals. The problem arises from the following mathematical result, valid for arbitrary functions and variables [253, p. 44]:

$$\frac{\partial}{\partial \alpha} \int_{x=f(\alpha)}^{x=g(\alpha)} F(\alpha, x) dx = \int_{x=f(\alpha)}^{x=g(\alpha)} \frac{\partial F(\alpha, x)}{\partial \alpha} dx$$

Next he goes to his main calculation. He replaces the straight cylindrical conductor with a toroidal one, like a ring conducting an azimuthal current. He calls the greater radius of the ring r and its smaller radius α, supposing $\alpha \ll r$. He considers the electrostatic potential null at the azimuth angle $\psi = \pi$ rad and discontinuous at $\psi = 0$ rad. See Figure A.5.

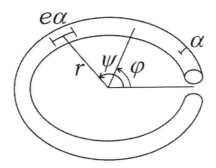

Figure A.5: Conducting ring with greater radius r and smaller radius α, with $\alpha \ll r$. The potential is discontinuous at $\psi = 0$ rad and null at $\psi = \pi$ rad. Weber wants to calculate the tangential electric field, E_ψ, at the angle ψ, $E_\psi(\psi)$. The angle φ is the variable angle of integration.

That is, the potential $F(\psi)$ [represented by Weber as $F\psi$] is such that $F(0) = -F(2\pi)$.[15] It is then given by $F(\psi) = c(\psi - \pi)$, where $[F(2\pi) - F(0)]/2\pi r = c/r$ is the value of the tangential electromotive force [our electric field] assumed constant along the ring. He utilizes his linear approach in order to calculate, in

$$+ \left\{ \frac{\partial g(\alpha)}{\partial \alpha} F[\alpha, g(\alpha)] - \frac{\partial f(\alpha)}{\partial \alpha} F[\alpha, f(\alpha)] \right\} . \tag{A.7}$$

In order to arrive at Eq. (A.3) beginning with Eq. (A.5) we would need to utilize the following expression (obtained from Eq. (A.7)):

$$E_{x'}(x') = -\frac{\partial \Phi(x')}{\partial x'} + \frac{\partial (x' - e\alpha/2)}{\partial x'} \frac{2\pi\alpha[a + b(x' - e\alpha/2)]}{\sqrt{[x' - (x' - e\alpha/2)]^2}}$$

$$- \frac{\partial (x' + e\alpha/2)}{\partial x'} \frac{2\pi\alpha[a + b(x' + e\alpha/2)]}{\sqrt{[x' - (x' + e\alpha/2)]^2}}$$

$$= -\frac{\partial \Phi(x')}{\partial x'} - 4\pi\alpha b . \tag{A.8}$$

And this equation coincides with Eq. (A.3) if we utilize Eq. (A.6). This is the correct approach if we wish to obtain the electric field $E(x')$ utilizing Weber's approximate method and beginning with an equivalent to a potential function. That is, we need to follow this approach if we begin with the function $\Phi(x')$ and wish to obtain $E(x')$ by differentiation. We will return to this point when considering Weber's next calculation.

[15] At $\psi = 0$ rad there should be the point of contact between copper and zinc, or a chemical battery, or another non-electrostatic source of electromotive force. Weber's α, r and F are equivalent to our r_0, R_0 and ϕ, respectively. See Fig. 13.1.

a general way, the electrostatic potential at the angle ψ along the ring due to all surface charges, integrating from from $\varphi = \psi + e\alpha/2r$ to $\varphi = 2\pi + \psi - e\alpha/2r$.[16]

He calls $f\varphi d\varphi$ the amount of free electricity in the arc element $rd\varphi$, where $f\varphi$ is the angular density of free charge (it is not yet specified whether this charge density is a linear function of the angle φ). That is, $f\varphi$ is the angular density of free electricity along the ring as a function of the azimuth angle φ. [From now on we will call it $f(\varphi)$. That is, $f(\varphi)$ has units of charge per angle, or Coulomb per radian in the SI.] He mentions that according to Ohm's hypothesis, the density of charges along a uniformly resistive conductor should be a linear function of the length along the circuit.[17] As we have a ring this would imply, according to Weber, that the angular density of charges should be given by $f(\varphi) = a(\varphi - \pi)$.[18] Weber then decides to test if this linear hypothesis is valid for a ring.

Instead of calculating the electric field directly, as he had done in the case of a linear conductor, he decided to calculate the electrostatic potential.[19] To this end he divides the circumference of radius r into two parts, ABD and DCA. See Figure A.6.

The points A, B, D and C are located at $\varphi = 0$ rad, $\varphi = \psi$ (where he wants

---

[16] That is, instead of performing an integration over the surface of the ring, he performs only a linear integration replacing the ring by a circumference with an appropriate linear charge density. He calculates the potential at the angle ψ, where the test charge will be located. His integration can be thought of as going from $\varphi = 0$ rad to $\varphi = \pi$ rad, except for the region between $\psi - e\alpha/2r$ and $\psi + e\alpha/2r$.

[17] This can be seen in p. 456 of Ohm's work [252].

[18] It should be observed that the a here has no relation with the a of the previous surface charge density of a cylinder given by $a + bx$. Weber's approach is analogous to the one utilized in Chapter 6. That is, he supposes a distribution of source charges and from them calculate the potential and electric field. The approach utilized in Chapter 13 was the opposite. In Chapter 13 it was given the potential along the surface of the conductor, Laplace's equation was solved, yielding the potential everywhere in space. Then the electric field was obtained as minus the gradient of the potential. And finally the surface charges were obtained by applying Gauss's law at the interface between the conductor and the external medium. For the ring we obtained a density of surface charges given by Eq. (13.63). Far from the battery this is reduced to Eq. (13.65), namely, $\sigma(\varphi) = \sigma_A + \sigma_B \varphi/2\pi$. Far from the battery the linear charge density which we obtained was given by Eq. (13.74). Weber's angular density of charges, $f(\varphi)$, is given by R_0 times the linear charge density. That is (far from the battery and utilizing $\eta_0 \gg 1$, $a \approx R_0$, $a/\sinh\eta_0 = r_0$ and $\cosh\eta_0 = R_0/r_0$):

$$f(\varphi) = R_0\left(\lambda_A + \lambda_B \frac{\varphi}{2\pi}\right) = 2\pi r_0 R_0 \left(\sigma_A + \sigma_B \frac{\varphi}{2\pi}\right)$$
$$= 2\pi R_0 \varepsilon_0 \left[\frac{\phi_A}{\ln(8R_0/r_0)} + g(\eta_0)\phi_B \frac{\varphi}{2\pi}\right]. \quad (A.9)$$

Comparing this expression with Weber's expression, $f(\varphi) = a(\varphi - \pi)$, we find that Weber's a is equivalent to our $r_0 R_0 \sigma_B = R_0 \varepsilon_0 g(\eta_0)\phi_B$.

[19] In principle he would need to integrate

$$\Phi(\psi) \equiv \left[\int_{\varphi=0}^{\psi-e\alpha/2r} + \int_{\varphi=\psi+e\alpha/2r}^{2\pi}\right] \frac{a(\varphi - \pi)d\varphi}{r\sqrt{2}\sqrt{1-\cos(\psi-\varphi)}}. \quad (A.10)$$

However, if he tried to perform this direct integration he would end up needing to evaluate $\int xdx/\sin x$. The solution of this indefinite integral yields an infinite series, namely [203, p. 233]:

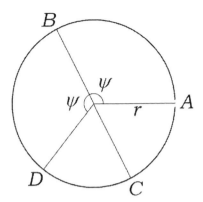

Figure A.6: Weber's configuration to integrate the potential.

to know the value of the potential), $\varphi = 2\psi$ and $\varphi = \psi + \pi$, respectively. In order to calculate the potential at B due to the charges spread along the arc ABD, with the exception of the small arc $e\alpha/r$ around B, he considers two charge elements symmetrically located around B, at angles $\pm \chi$ from B. The charge elements located at $\varphi = \psi \pm \chi$, in elementary arcs of length $rd\chi$, are given by $a(\psi \pm \chi - \pi)d\chi$. Each of these charge elements is at the same distance $2r\sin(\chi/2)$ from B. By adding the contributions of these two charge elements he obtains the differential potential at B as given by $a(\psi - \pi)d\chi/r\sin(\chi/2)$. This was a very good idea in order to avoid the integral of $xdx/\sin x$. After integration he obtains the potential

$$\frac{a(\psi - \pi)}{r} \int_{\chi = \frac{e\alpha}{2r}}^{\psi} \frac{d\chi}{\sin\frac{\chi}{2}} = \frac{2a(\psi - \pi)}{r} \left(\log \tan \frac{\psi}{4} - \log \tan \frac{e\alpha}{8r} \right). \quad (A.12)$$

To obtain the potential at B due to the charges located around the arc DCA he proceeds in a similar way. He considers two charge elements located symmetrically around C, at angles $\pm \chi$ from C. Relative to A these two charge elements are located at $\varphi = \psi + \pi \pm \chi$. Both are at the same distance $2r\sin[(\pi - \psi)/2]$ from B. The potential due to the sum of these two charge elements calculated at B is then given by $a\psi d\chi/r\cos(\chi/2)$. After integration he obtains

$$\frac{a\psi}{r} \int_{\chi=0}^{\pi-\psi} \frac{d\chi}{\cos\frac{\chi}{2}} = -\frac{2a\psi}{r} \log \tan \frac{\psi}{4}. \quad (A.13)$$

$$\int \frac{xdx}{\sin x} = x + \sum_{k=1}^{\infty} (-1)^{k+1} \frac{2(2^{2k-1} - 1)}{(2k+1)!} B_{2k} x^{2k+1}. \quad (A.11)$$

It is difficult to put this infinite series in closed form.

Instead of solving this integral directly, Weber utilizes an ingenious approach by taking advantage of the symmetrical distribution of charges along the circumference, as we will show below. In this way he avoids this integral.

By adding Eqs. (A.12) and (A.13) he obtains the total potential at $\varphi = \psi$ as given by[20]

$$-\frac{2a\psi}{r}\log\tan\frac{e\alpha}{8r} - \frac{2a\pi}{r}\left(\log\tan\frac{\psi}{4} - \log\tan\frac{e\alpha}{8r}\right). \quad (A.15)$$

By making the derivative of this expression with respect to the arc $r\psi$, namely, $d/rd\psi$, Weber obtains the following expression for the magnitude of the tangential component of the electromotive force [the absolute value of our electric field] at the angle ψ due to all surface charges along the ring, except for the charges in the arc $e\alpha/r$ around ψ:

$$-\frac{2a}{r^2}\log\tan\frac{e\alpha}{8r} - \frac{a\pi}{r^2\sin(\psi/2)}. \quad (A.16)$$

That is, Weber initially calculated the potential at the angle ψ as given by Eq. (A.14). He then obtained absolute value of the tangential component of the

---

[20] This final value obtained by Weber can be written as

$$\Phi(\psi) \equiv \left[\int_{\varphi=0}^{\psi-e\alpha/2r} + \int_{\varphi=\psi+e\alpha/2r}^{2\pi}\right]\frac{a(\varphi-\pi)d\varphi}{r\sqrt{2}\sqrt{1-\cos(\psi-\varphi)}}$$

$$= -\frac{2a\psi}{r}\log\tan\frac{e\alpha}{8r} - \frac{2a\pi}{r}\left(\log\tan\frac{\psi}{4} - \log\tan\frac{e\alpha}{8r}\right). \quad (A.14)$$

electric field at ψ as given by $E_\psi(\psi) = d\Psi_\psi(\psi)/rd\psi$, obtaining Eq. (A.16).[21]

He then mentions that this value is approximately constant only for $\psi \approx \pi$ rad, i.e., far from the point of discontinuity in the potential [far from the battery]. When we are close to $\psi = 0$ rad or to $\psi = \pi$ rad, the magnitude of this longitudinal electromotive force is smaller than its magnitude at $\psi = \pi$ rad. He concludes that Ohm's hypothesis is only valid for the middle part of the circuit (that is, for $\psi \approx \pi$ rad).

He mentions that it is then also necessary to consider the charges which are located in the cross-sections of the ring where there is a discontinuity in the potential, charges which had not been considered by Ohm. He calls $\pm\varepsilon$ the amount of these opposite surface charges (which he considers in his simplified model as concentrated at points) and δ the small distance separating them. See Figure A.7.

After calculating the absolute value of the longitudinal electromotive force [that is, the tangential electric field along the ring] acting at the angle ψ due to this dipole, Weber obtains the result[22]

---

[21] Weber could have obtained the tangential component of the electric field by direct integration. That is,

$$E_\psi(\psi) = \left(\int_{\varphi=0}^{\psi-e\alpha/2r} + \int_{\varphi=\psi+e\alpha/2r}^{2\pi} \right) \frac{a(\varphi-\pi)\sin(\psi-\varphi)d\varphi}{\sqrt{8}r^2[1-\cos(\psi-\varphi)]^{3/2}}$$

$$= \frac{2a}{r^2}\ln\tan\frac{e\alpha}{8r} + \frac{a\pi}{r^2\sin(\psi/2)} - \frac{e\alpha a}{2r^3} \frac{1}{\sin\frac{e\alpha}{4r}} . \quad (A.17)$$

The last term on the right hand side does not appear in Weber's expression, Eq. (A.16). This is due to the same problem discussed in footnote 14.

In order to arrive at Eq. (A.17) beginning with Eq. (A.14) and taking into account Eq. (A.7), Weber should have utilized:

$$E_\psi(\psi) = -\frac{1}{r}\frac{\partial \Phi}{\partial \psi} + \frac{\partial(\psi - e\alpha/2r)}{\partial \psi} \frac{a(\psi - e\alpha/2r - \pi)}{r^2\sqrt{2}\sqrt{1-\cos(\psi - (\psi - e\alpha/2r))}}$$

$$- \frac{\partial(\psi + e\alpha/2r)}{\partial \psi} \frac{a(\psi + e\alpha/2r - \pi)}{r^2\sqrt{2}\sqrt{1-\cos(\psi - (\psi + e\alpha/2r))}}$$

$$= -\frac{1}{r}\frac{\partial \Phi}{\partial \psi} - \frac{e\alpha a}{2r^3 \sin\frac{e\alpha}{4r}} . \quad (A.18)$$

And this coincides with Eq. (A.17) based on Eq. (A.14).

At $\psi = \pi$ rad and with $\alpha \ll r$ we obtain from Eq. (A.17):

$$E_\psi(\psi = \pi \text{ rad}) \approx \frac{2a}{r^2}\left(\ln\frac{\alpha}{8r} + \frac{\pi}{2}\right) . \quad (A.19)$$

Utilizing Weber's a as our $r_0 R_0 \sigma_B$ (as we saw in footnote 18) and also his α and r as our r_0 and R_0, respectively, the latter equation can be written as (dividing the right hand side by $4\pi\varepsilon_0$ in order to obtain the electric field in the international system of units):

$$E_\psi(\psi = \pi \text{ rad}) \approx -\frac{r_0 \sigma_B}{2\pi\varepsilon_0 R_0}\left(\ln\frac{8R_0}{r_0} - \frac{\pi}{2}\right) . \quad (A.20)$$

[22] We have checked this result and it is correct. There should be an overall minus sign in front of this expression if we wish to express the algebraic value of the tangential electric field due to this dipole.

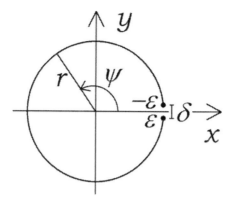

Figure A.7: Simplified model proposed by Weber to consider the opposite surface charges $\pm\varepsilon$ in the discontinuity of the potential, separated by a small distance δ.

$$\frac{1+\cos^2(\psi/2)}{\sin^3(\psi/2)}\frac{\delta\varepsilon}{8r^3}. \qquad (A.21)$$

To find $\delta\varepsilon$ he considers the value of the net electromotive force [due to the surface charges along the ring and due to the dipole at the discontinuity]. He specifies that this net electromotive force should be almost constant with ψ. That is, he chooses $\delta\varepsilon$ such that the second and third derivatives of the electromotive force with regard to the azimuthal angle ψ go to zero at $\psi = \pi$ rad. With this condition he obtains[23]

$$\delta\varepsilon = \frac{8a\pi r}{5}. \qquad (A.22)$$

Combining Eqs. (A.21) and (A.22) with the previous result arising from Ohm's linear hypothesis yields as the final result:[24]

$$\frac{2a}{r^2}\log\cot\frac{e\alpha}{8r} + \frac{2a\pi}{5r^2\sin^3(\psi/2)}\left(3\cos^2\frac{\psi}{2} - 2\right). \qquad (A.24)$$

In this case the absolute value of the electromotive force for $\psi \neq \pi$ rad is greater than its absolute value at $\psi = \pi$ rad, while with only Ohm's linear hypothesis he had found that the electromotive force was smaller for $\psi \neq \pi$

---

[23]This result is also correct.
[24]Combining Eq. (A.22) with the negative of Eq. (A.21), as we discussed in footnote 22, together with Eq. (A.17), we obtain:

$$E_\psi(\psi) = \frac{2a}{r^2}\ln\tan\frac{e\alpha}{8r} + \frac{a\pi}{r^2\sin(\psi/2)} - \frac{e\alpha a}{2r^3}\frac{1}{\sin\frac{e\alpha}{4r}} - \frac{a\pi}{5r^2}\frac{1+\cos^2(\psi/2)}{\sin^3(\psi/2)}. \qquad (A.23)$$

rad than at $\psi = \pi$ rad. He then concluded [32] [38, p. 382] (our words in square brackets): "The correct hypothesis about the distribution of the free electricity, from which should result an equal electromotive force [tangential or longitudinal electric field] in all parts [along the ring] is then contained between both hypotheses above, which means the same as: the [surface] electric charge of the circuit increases from the neutral point [$\psi = \pi$ rad, opposite to the battery] to the contact point [$\psi = 0$ rad, where there is contact between copper and zinc, or the chemical battery, or another non-electrostatic source of electromotive force] not uniformly, but accelerates gradually."[25] He goes on to write: "The everywhere equal electromotive force which follows from this [analysis] will be situated presumably between the two limiting values given by the hypotheses above, namely

$$\frac{2a}{r^2}\left(\log\cot\frac{e\alpha}{8r} - \frac{\pi}{2}\right) \tag{A.25}$$

and

$$\frac{2a}{r^2}\left(\log\cot\frac{e\alpha}{8r} - \frac{2\pi}{5}\right). \tag{A.26}$$

The factor a is related to the *slope* of the [surface] electric charge in the middle of the circuit [$\psi = \pi$ rad], when slope is understood, according to Ohm, as the differential quotient of the charge $f\varphi$ [that is, charge per angle $f(\varphi)$] in relation to the arc φ [in other words, $a = df/d\varphi$]."

In Section 31 Weber presents a mathematical method to estimate the distribution of surface charges in a linear conductor (that is, a filiform conductor which can be straight or curved) carrying a steady current, in different cases. His method can also yield an estimation of the magnitude of the corresponding electric field inside the conductor produced by this distribution of surface charges.

Section 32 is called "Proof of how a necessary distribution of free electricity on the surface of a closed conductor arises when it carries a steady and uniform current." He considers a closed circuit with only one point acted upon by an electromotive force [like the contact of copper and zinc]. Only the charges in this point will begin to move, but according to Weber, this will cause a distribution of free charges along the whole conductor. And there will be a specific distribution of free charges which will create an electromotive force [electric field] at all other points of the circuit, allowing it to carry a steady current. He goes on to mention that this distribution of surface charges does not produce electrostatic

---

[25]That is, he concluded that the surface charge density along the resistive ring carrying a steady current grows linearly with the azimuthal angle ψ only close to $\psi = \pi$ rad, i.e., opposite to the battery. When we approach the battery the density of surface charges must grow faster than linearly with the azimuthal angle ψ, in order to produce a uniform tangential electric field at all points along the ring. That is, the surface charge density cannot increase as a function of ψ simply as $\sigma = C_1 + C_2\psi$. If it did increase linearly with ψ, the magnitude of the tangential electric field would not be constant at all points along the ring. This is a remarkable prediction confirmed by our calculations in Chapter 13. See specially Figure 13.7.

equilibrium, otherwise the net electric force at any point along the surface of the conductor would be orthogonal to the conductor. He says that this distribution of surface charges will create both a normal component of the electric force at the surface [of the conductor], and a tangential component. This means, in Weber's view, that the free charges along the surface of the conductor carrying a steady current cannot be stationary, but must participate in the motion of the internal current. But he also shows that the motion does not imply a temporal variation of this distribution of surface charges. That is, this distribution will not change with time for steady currents, as at any section along the surface there will be an equal amount of charges entering and leaving the section. [The density of surface charges will then be a function of the longitudinal coordinate only, and not a function of time.]

In Section 33 he mentions that during the printing of his work, Kirchhoff's paper dealing with the same subject was published [24] (this paper has been translated into English [27]). We discuss this paper in the next Appendix. Weber quotes the final section of Kirchhoff's paper. This Section of Weber's work is important to indicate that Weber and Kirchhoff arrived at essentially the same ideas independently of one another, both trying to improve upon Ohm's work and hypotheses. But Weber was the only one who attempted to calculate explicitly the distribution of surface charges in specific configurations.

Section 34 is called "To determine, through a comparison of electromotive and galvanometric observations of a galvanic circuit, the relative velocity between two electrical masses in which no attraction nor repulsion arises." Weber derives here a theoretical relation of the fundamental constant which appears in his law of force (1846), with the current, resistance and electromotive force in a circuit carrying a steady current. According to Weber's force law, Eq. (1.1), when two charges approach or separate from from one another with a constant relative velocity $\dot{r} = \sqrt{2}c$ (with the modern nomenclature that $c = 3 \times 10^8$ m/s), they will not affect one another, regardless of the signs of the charges. That is, the Coulombian component of the force will be balanced by the velocity component, yielding zero net force between them. Only in 1855-56 did Weber and Kohlrasch succeed in obtaining experimentally the value of this fundamental constant. See Section 1.4 for references.

Section 35 is called "On the ratio of the velocity of the flow to the velocity of the propagation of the current." Here Weber presents a first theoretical comparison of the drifting velocity of charges in a conductor carrying a constant current, with the velocity for the propagation of a variable current along this conductor. The numerical values of these two velocities were not yet known at that time, as no experiments had given their orders of magnitude.

Section 36 is called "On the origin of the resistance of conductors." He begins by mentioning that for a complete understanding of the resistance it is not enough to define it by its effect (as the ratio between electromotive force and current given by Ohm's law). That is, it is also necessary to define resistance by its origin. In particular we need to know if it comes through the ponderable part of the current or from its electric fluid. Weber asks: What is the origin of the force that creates resistance to the motion of the charges against the electromotive force accelerating them? He wants to know if this force is purely electric, or if it acts upon the ponderable particles of the current (due to forces having another origin, like molecular forces). In his reasoning, he considers initially Fechner's hypothesis, *i.e.*, he assumes a double current with an equal amount of positive and negative charges moving relative to the wire with equal and opposite velocities. He analyzes whether the encounter of these opposite charges might give rise to the resistive force, due only to electromagnetic forces between these charges. To this end he considers a simplified model in which only the negative charges move relative to the wire, while the positive charges remain fixed in the lattice. He is here departing from Fecher's hypothesis and coming close to the modern model of a current in metallic conductors in which only the electrons move relative to the lattice. But at that time no one knew about the existence of electrons and they also did not know the order of magnitude of the drift velocity of the mobile charges. Weber here imagines a negative charge making a Keplerian elliptical orbit around a positive charge due to a central force which falls as $1/r^2$, disregarding the components of his fundamental force law (1.1) which depend on the relative velocity and relative acceleration between the charges (by considering that these components have a small value in comparison to the greater value of the Coulombian component). When there is an electromotive force [like an external electric field] acting along the wire, it will perturb this orbit into a spiral form. The loops of this spiral will increase until the negative charges come into the sphere of action of another positive charge along the wire. It will orbit this second positive charge until it comes into the sphere of action of the third positive charge along the line composing the wire. This transference of the negative charge to the following positive charges will continue as long as the electromotive force acts upon the conductor. In the event this electromotive force stops acting, the negative charge will no longer move forward, but will continue to circle the specific positive charge around which it was moving when the electromotive force was interrupted. He concludes the Section by mentioning that it would be important to calculate the time interval needed by the negative charge to move in its spiral orbit from one positive charge to the next, but that this calculation should be difficult, as is shown by the perturbation theory of astronomy.[26].

This fifth part of Weber's paper is extremely important. Here we can see that he is one of the pioneers who pointed out the surface charges in resistive conductors carrying steady currents. The chemical battery or contact between two different metals, like copper and zinc, creates a difference of potential between two points. But what creates the uniform electric field tangential to the circuit at every point inside a resistive wire is the distribution of free charges along

---

[26]Weber's idea that the resistive force might be due to a newtonian central force falling as $1/r^2$ does not seem feasible to us for two main reasons. (1) The newtonian forces are conservative and (2) do not depend on the velocities of the interacting bodies. The resistive force responsible for Ohm's law, on the other hand, is non-conservative and proportional to the drifting velocities of the mobile charges, acting against the motion of these charges. The origin of this force must be sought somewhere else. The origin of these resistive forces is a very difficult topic in physics, and even today there is no clear answer to this question.

the surface of this wire. He correctly pointed out that these surface charges must be in motion together with the current, as the tangential electric field will act not only inside it, but also along the surface of the conductor. Moreover, he was probably the first to try to calculate this distribution of surface charges explicitly in a specific example. In particular he considered a ring of finite cross-section, much smaller than the length of the ring, with a small gap at one point where a non-electrostatic electromotive force acts. With an ingenious calculation he showed that the distribution of surface charges increases linearly with the azimuthal angle only in the region opposite to the battery. He showed that as we approach the gap the surface charge density must increase faster than linearly with the azimuthal angle, a remarkable result confirmed 150 years later when this problem was completely solved analytically, as described in Chapter 13 of this book. Weber goes even further, trying to understand the origin of the resistive force in terms of microscopic forces of electromagnetic origin between the interacting charges composing the current. This is a remarkable piece of work which deserves to be more widely known.

Weber produced another very important study in 1864 which continues the study of surface charges: "Electrodynamic measurements relating specially to electric oscillations" [254]. This is the fifth work in the series of "Electrodynamic measurements." To the best of our knowledge, it has also never been translated into English. The main theoretical derivations of this paper were obtained in 1857 or prior to that, but were not published at this time. A similar treatment was first published by Kirchhoff in 1857. As we discuss Kirchhoff's papers in the next Appendix, we will not enter into details here of Weber's similar findings which were delayed in publication. Kirchhoff's paper was published in *Poggendorff's Annalen*, now known as *Annalen der Physik*. Poggendorff wrote a note after Kirchhoff's paper relating that after seeing it he had occasion to meet Weber in Berlin. Weber showed him the paper he intended to publish, with essentially the same results as Kirchhoff's. But Weber had not yet sent it to print, as he was waiting for results of experiments on this topic to be performed together with R. Kohlrasch [255]. This paper by Weber was published in 1864. It deals with the propagation of electromagnetic signals along wires, taking into consideration variable currents and the effects of all surface charges upon the current. As we will see, Weber and Kirchhoff arrived at the telegraphy equation.

Appendix B

Gustav Kirchhoff and Surface Charges

Here we discuss three papers by Kirchhoff, one from 1849 and two from 1857 [24, 25, 26]. All of these papers have been translated into English [27, 28, 29]. For this reason we present only brief summaries of them.

In the first paper he pointed out a mistake in Ohm's hypothesis according to which a uniform volume density of electricity could remain at rest inside a conductor. Ohm assumed also that the electroscopic or electromotive forces acting along a resistive conductor carrying a steady current would be proportional to the variation of this volume density of charges as regards the longitudinal coordinate (in the case of a linear conductor). According to Kirchhoff, on the other hand, what is constant inside a conductor in electrostatic equilibrium is its electric potential, but not its volume charge density. The electromotive force inside a resistive conductor carrying a steady current is proportional to the variation of this potential with the longitudinal coordinate. And the potential itself originates from free charges spread along the surface of the conductor. Kirchhoff shows that even in the case of steady currents the potential will satisfy Laplace's equation inside the conductor. He does not try to calculate the distribution of these surface charges in any specific example. At the end of this first paper he shows that Weber's law of force between point charges is also compatible with Ohm's law, and with his reasoning of free charges along the surface of resistive conductors.

In his first paper of 1857 Kirchhoff derives the telegraphy equation for a signal propagating along a thin conducting wire. We present here his main results in vectorial notation and in the International System of Units SI, following a paper of 1999 [256]. Weber's simultaneous and more thorough work was delayed in publication, and was published only in 1864. Both worked independently of one another and predicted the existence of periodic modes of oscillation of the electric current propagating at light velocity in a conducting circuit of negligible resistance.

In his first paper of 1857, Kirchhoff considered a conducting circuit of circular cross-section which might be open or closed. Kirchhoff's wire could be straight or curved, provided the following assumption was satisfied: "that the form of the central line of the wire is such, that the distance between two of its points, between which a finite portion of the wire lies, is never infinitely small. By this supposition the case is excluded, that induction spirals are contained in the circuit." He wrote Ohm's law taking into account the free electricity along the surface of the wire and the induction due to the alteration of the strength of the current in all parts of the wire:

$$\vec{J} = -g\left(\nabla\phi + \frac{\partial \vec{A}}{\partial t}\right) \ . \tag{B.1}$$

Here $\vec{J}$ is the current density, g is the conductivity of the wire, ϕ is the electric potential and $\vec{A}$ is a function analogous to the modern magnetic vector potential (which Kirchhoff will calculate from Weber's force). He calculates ϕ by integrating the effect of all free surface charges:

$$\phi(x,y,z,t) = \frac{1}{4\pi\varepsilon_0} \int\int \frac{\sigma(x',y',z',t)da'}{|\vec{r}-\vec{r}\,'|} \ . \tag{B.2}$$

Here $\vec{r} = x\hat{x} + y\hat{y} + z\hat{z}$ is the point where the potential is being calculated, t is the time, and σ is the free surface charge. Kirchhoff then performed a remarkable calculation, integrating this equation over the whole surface of the wire of length ℓ and radius a without specifying the behaviour of σ with regard to the variables x', y', z' or t, but only the requirement that $a \ll \ell$. Moreover, he supposed that the current density was the same at all points of the periphery of a cross-section in the wire (that is, he neglected the effects of curvatures in the wire) and that it was never infinitely large. With only these assumptions he arrived finally at:

$$\phi(s,t) = \frac{a\sigma(s,t)}{\varepsilon_0}\ln\frac{\ell}{a} \ . \tag{B.3}$$

Here s is a variable distance along the wire from a fixed origin. See Figure B.1. This is equivalent to our Eq. (6.8). While our equation was derived for a straight wire carrying a steady current, Kirchhoff obtained it for a wire which might be straight or slightly curved. Moreover, in his calculation the potential, current and surface charge density could also be a function of time. This was a remarkable result, also obtained by Weber and published in 1864 [254].

He obtains the vector potential $\vec{A}$ from Weber's force, Eq. (1.1). That is, the component of this force which depends upon the acceleration of the charges can be written as $-q\partial\vec{A}/\partial t$, with a vector potential given by

$$\vec{A}(x,y,z,t) = \frac{\mu_0}{4\pi}\int\int\int\left[\vec{J}(x',y',z',t)\cdot(\vec{r}-\vec{r}\,')\right](\vec{r}-\vec{r}\,')\frac{dx'dy'dz'}{|\vec{r}-\vec{r}\,'|^3} \ . \tag{B.4}$$

Here the integration is through the volume of the wire.

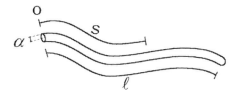

Figure B.1: A long curved conductor of length ℓ and radius α. The variable s represents a distance along the wire from a fixed origin O.

After integrating this expression he arrived at

$$\vec{A}(s,t) = \frac{\mu_0}{2\pi} I(s,t) \left(\ln \frac{\ell}{\alpha} \right) \hat{s} , \tag{B.5}$$

where $I(s,t)$ is the variable current.

Given that $I = J\pi\alpha^2$ and that $R = \ell/(\pi g \alpha^2)$ is the resistance of the wire, the longitudinal component of Ohm's law could then be written as

$$\frac{\partial \sigma}{\partial s} + \frac{1}{2\pi\alpha} \frac{1}{c^2} \frac{\partial I}{\partial t} = -\frac{\varepsilon_0 R}{\alpha \ell \ln(\ell/\alpha)} I . \tag{B.6}$$

In order to relate the two unknowns, σ and I, Kirchhoff utilized the equation for the conservation of charges, which he wrote as

$$\frac{\partial I}{\partial s} = -2\pi\alpha \frac{\partial \sigma}{\partial t} . \tag{B.7}$$

To the best of our knowledge this was the first time that this fundamental equation for the conservation of charges was published in the literature.

When these two relations are equated, they yield the equation of telegraphy, namely:

$$\frac{\partial^2 \xi}{\partial s^2} - \frac{1}{c^2} \frac{\partial^2 \xi}{\partial t^2} = \frac{2\pi\varepsilon_0 R}{\ell \ln(\ell/\alpha)} \frac{\partial \xi}{\partial t} , \tag{B.8}$$

where ξ can represent I, σ, ϕ or the longitudinal component of $\vec{A}$.

If the resistance is negligible, this equation predicts the propagation of signals along the wire with light velocity. As Kirchhoff put it, the velocity of propagation of an electric wave "is independent of the cross-section, of the conductivity of the wire, also, finally, of the density of the electricity: its value is 41950 German miles in a second, hence very nearly equal to the velocity of light in vacuo."

Equations similar to Eq. (B.8) can be found in pages 123 and 125 of the second part of Volume 4 of Weber's Collected Papers (original paper of 1864) [254, 39].

In his second paper of 1857 Kirchhoff generalizes this first work in order to consider three-dimensional conductors of arbitrary shape. We discussed this briefly in 1994 [29]. The results are essentially the same as before, but now he shows that it is possible to have free electricity distributed throughout the substance of the conductor, in the case of a current varying in time and in space.

Recently we developed Kirchhoff's ideas in the international system of units and applied them to the propagation of electromagnetic signals in a coaxial cable, a situation which was not considered by Kirchhoff [30, 31, 257, 258].

It should be stressed that the works of Kirchhoff of 1857 were published before Maxwell wrote down his equations in 1861-64, establishing the electromagnetic theory of light. When Maxwell introduced the displacement current $(1/c^2)\partial \vec{E}/\partial t$ he was utilizing Weber's constant c. He was also aware of Weber and Kohlrasch's measurement of 1854-56 that c had the same value as light velocity in vacuum. He also knew Kirchhoff's derivation of the telegraphy equation yielding the propagation of electromagnetic signals at light velocity. Kirchhoff's work of 1857 was quoted only once by Maxwell in the note to paragraph 805, p. 450 of his *Treatise* [158]. It should be remarked that this specific citation does not appear in the Index at the end of Maxwell's *Treatise*. For this reason it may not have been noted by some authors. In note 26 of Schaffer's paper we find the following important remark regarding Maxwell's knowledge of Kirchhoff's first paper of 1857: "In the early 1870s Maxwell made detailed notes on Kirchhoff's paper on electricity in wires: see Cambridge University Library MSS ADD 7655 Vn/1, p. 44 ff" [259].

As mentioned above, Kirchhoff's paper of 1857 was published earlier than Weber's paper of 1864, although both of them arrived at essentially the same results independently of one another and at the same time, as pointed out by Poggendorff, the editor of the *Annalen der Physik* (at that time called *Poggendorff's Annalen*) in 1857 [255].

Bibliography

[1] A. K. T. Assis, W. A. Rodrigues Jr., and A. J. Mania. The electric field outside a stationary resistive wire carrying a constant current. *Foundations of Physics*, 29:729–753, 1999.

[2] R. Clausius. Ableitung eines neuen elektrodynamischen Grundgesetzes. *Journal für reine und angewandte Mathematik*, 82:85–130, 1877.

[3] A. O'Rahilly. *Electromagnetic Theory — A Critical Examination of Fundamentals*. Dover, New York, 1965.

[4] U. Bartocci and M. M. Capria. Some remarks on classical electromagnetism and the principle of relativity. *American Journal of Physics*, 59:1030–1032, 1991.

[5] U. Bartocci and M. M. Capria. Symmetries and asymmetries in classical and relativistic electrodynamics. *Foundations of Physics*, 21:787–801, 1991.

[6] F. Cardone and R. Mignani. Observation of a DC voltage induced by a steady magnetic field: A possible electromagnetic breakdown of Lorentz invariance? *Physics Essays*, 13:645–656, 2000.

[7] U. Bartocci, F. Cardone, and R. Mignani. Looking for a possible breakdown of local Lorentz invariance for electromagnetic phenomena: theory and first experimental results. *Foundations of Physics Letters*, 14:51–64, 2001.

[8] J. R. Reitz, F. J. Milford, and R. W. Christy. *Foundations of Electromagnetic Theory*. Addison-Wesley, Reading, 4th edition, 1993.

[9] B. R. Russell. Surface charges on conductors carrying steady currents. *American Journal of Physics*, 36:527–529, 1968.

[10] J. D. Jackson. *Classical Electrodynamics*. John Wiley & Sons, New York, 1962.

[11] J. D. Jackson. *Classical Electrodynamics*. John Wiley & Sons, New York, second edition, 1975.

[12] J. D. Jackson. Surface charges on circuit wires and resistors play three roles. *American Journal of Physics*, 64:855–870, 1996.

[13] J. D. Jackson. *Classical Electrodynamics*. John Wiley & Sons, New York, third edition, 1999.

[14] W. F. Edwards. Inertia and an alternative approach to the theory of interactions. *Proceedings of the Utah Academy of Science, Arts, and Letters*, 51, Part 2:1–7, 1974.

[15] W. F. Edwards, C. S. Kenyon, and D. K. Lemon. Continuing investigation into possible electric fields arising from steady conduction currents. *Physical Review D*, 14:922–938, 1976.

[16] D. J. Griffiths. *Introduction to Electrodynamics*. Prentice Hall, Englewood Cliffs, second edition, 1989.

[17] C. A. Coombes and H. Laue. Electric fields and charge distributions associated with steady currents. *American Journal of Physics*, 49:450–451, 1981.

[18] P. Lorrain, D. R. Corson, and F. Lorrain. *Electromagnetic Fields and Waves*. W. H. Freeman and Co., New York, 3rd edition, 1988.

[19] S. Seely. *Introduction to Electromagnetic Fields*. McGraw-Hill, New York, 1958.

[20] B. D. Popović. *Introductory Engineering Electromagnetics*. Addison-Wesley, Reading, 1971.

[21] A. K. T. Assis. Deriving Ampère's law from Weber's law. *Hadronic Journal*, 13:441–451, 1990.

[22] A. K. T. Assis. Can a steady current generate an electric field? *Physics Essays*, 4:109–114, 1991.

[23] A. K. T. Assis. *Weber's Electrodynamics*. Kluwer Academic Publishers, Dordrecht, 1994. ISBN: 0-7923-3137-0.

[24] G. Kirchhoff. Ueber eine Ableitung der Ohm'schen Gesetze, welche sich an die Theorie der Elektrostatik anschliesst. *Annalen der Physik*, 78:506–513, 1849. Reprinted in G. Kirchhoff's *Gesammelte Abhandlungen* (Barth, Leipzig, 1882), pp. 49-55.

[25] G. Kirchhoff. Ueber die Bewegung der Elektricität in Drähten. *Annalen der Physik*, 100:193–217, 1857. Reprinted in G. Kirchhoff's *Gesammelte Abhandlungen* (Barth, Leipzig, 1882), pp. 131-154.

[26] G. Kirchhoff. Ueber die Bewegung der Elektricität in Leitern. *Annalen der Physik*, 102:529–544, 1857. Reprinted in G. Kirchhoff's *Gesammelte Abhandlungen* (Barth, Leipzig, 1882), pp. 154-168.

[27] G. Kirchhoff. On a deduction of Ohm's law in connexion with the theory of electrostatics. *Philosophical Magazine*, 37:463–468, 1850.

[28] G. Kirchhoff. On the motion of electricity in wires. *Philosophical Magazine*, 13:393–412, 1857.

[29] P. Graneau and A. K. T. Assis. Kirchhoff on the motion of electricity in conductors. *Apeiron*, 19:19–25, 1994.

[30] A. K. T. Assis. On the propagation of electromagnetic signals in wires and coaxial cables according to Weber's electrodynamics. *Foundations of Physics*, 30:1107–1121, 2000.

[31] A. K. T. Assis and J. A. Hernandes. Telegraphy equation from Weber's electrodynamics. *IEEE Transactions on Systems and Circuits II*, 52:289–292, 2005.

[32] W. Weber. Elektrodynamische Maassbestimmungen insbesondere Widerstandsmessungen. *Abhandlungen der Königl. Sächs. Gesellschaft der Wissenschaften, mathematisch-physische Klasse*, 1:199–381, 1852. Reprinted in Wilhelm Weber's *Werke*, Vol. 3, H. Weber (ed.), (Springer, Berlin, 1893), pp. 301-471.

[33] R. P. Feynman, R. B. Leighton, and M. Sands. *The Feynman Lectures on Physics*. Addison-Wesley, Reading, 1964. Volume 2: Mainly Electromagnetism and Matter.

[34] E. M. Purcell. *Electricity and Magnetism*, volume 2 of *Berkeley Physics Course*. McGraw-Hill, New York, 1965.

[35] O. D. Jefimenko. Is magnetic field due to an electric current a relativistic effect? *European Journal of Physics*, 17:180–182, 1996.

[36] W. Weber. *Wilhelm Weber's Werke*, W. Voigt, (Ed.), volume 1, *Akustik, Mechanik, Optik und Wärmelehre*. Springer, Berlin, 1892.

[37] W. Weber. *Wilhelm Weber's Werke*, E. Riecke (Ed.), volume 2, *Magnetismus*. Springer, Berlin, 1892.

[38] W. Weber. *Wilhelm Weber's Werke*, H. Weber (Ed.), volume 3, *Galvanismus und Elektrodynamik*, first part. Springer, Berlin, 1893.

[39] W. Weber. *Wilhelm Weber's Werke*, H. Weber, (Ed.), volume 4, *Galvanismus und Elektrodynamik*, second part. Springer, Berlin, 1894.

[40] W. Weber and E. H. Weber. *Wilhelm Weber's Werke*, E. Riecke (Ed.), volume 5, *Wellenlehre auf Experimente gegründet oder über die Wellen tropfbarer Flüssigkeiten mit Anwendung auf die Schall- und Lichtwellen*. Springer, Berlin, 1893.

[41] W. Weber. *Wilhelm Weber's Werke,* F. Merkel and O. Fischer (Eds.), volume 6, *Mechanik der Menschlichen Gehwerkzeuge.* Springer, Berlin, 1894.

[42] W. Weber. On the theory of diamagnetism - Letter from Professor Weber to Prof. Tyndall. *Philosophical Magazine,* 10:407–410, 1855.

[43] W. Weber. On the measurement of electric resistance according to an absolute standard. *Philosophical Magazine,* 22:226–240 and 261–269, 1861.

[44] W. Weber. Electrodynamic measurements — Sixth memoir, relating specially to the principle of the conservation of energy. *Philosophical Magazine,* 43:1–20 and 119–149, 1872.

[45] C. F. Gauss and W. Weber. Results of the observations made by the Magnetic Association in the year 1836. In R. Taylor, editor, *Scientific Memoirs,* Vol. 2, pages 20–97, New York, 1966. Johnson Reprint Corporation.

[46] W. Weber. On the measurement of electro-dynamic forces. In R. Taylor, editor, *Scientific Memoirs,* Vol. 5, pages 489–529, New York, 1966. Johnson Reprint Corporation.

[47] W. Weber. On the excitation and action of diamagnetism according to the laws of induced currents. In R. Taylor, editor, *Scientific Memoirs,* Vol. 5, pages 477–488, New York, 1966. Johnson Reprint Corporation.

[48] W. Weber. On the connexion of diamagnetism with magnetism and electricity. In J. Tyndall and W. Francis, editors, *Scientific Memoirs,* Vol. 7, pages 163–199, New York, 1966. Johnson Reprint Corporation.

[49] W. Weber. Observations on the arrangment and use of the bifilar magnetometer. In R. Taylor, editor, *Scientific Memoirs,* Vol. 2, pages 268–280, New York, 1966. Johnson Reprint Corporation.

[50] W. Weber. On a transportable magnetometer. In R. Taylor, editor, *Scientific Memoirs,* Vol. 2, pages 565–586, New York, 1966. Johnson Reprint Corporation.

[51] W. Weber. An extract from remarks on the term-observations for 1839, of the German Magnetic Association. In R. Taylor, editor, *Scientific Memoirs,* Vol. 2, pages 587–588, New York, 1966. Johnson Reprint Corporation.

[52] W. Weber. Determinations of electrodynamic measure: particularly in respect to the connection of the fundamental laws of electricity with the law of gravitation. English translation by G. Gregory, unpublished, 21st Century Science and Technology, 1996.

[53] C. F. Gauss and W. E. Weber. Text of the Gauss-Weber correspondence. *21st Century*, 9(3):41–43, 1996. English translation by Susan P. Johnson.

[54] W. Weber. Aphorisms. *21st Century*, 10(2):50–53, 1997. English translation by J. Tennenbaum.

[55] W. Weber and R. Kohlrausch. On the amount of electricity which flows through the cross-section of the circuit in galvanic currents. In F. Bevilacqua and E. A. Giannetto, editors, *Volta and the History of Electricity*, pages 287–297. Università degli Studi di Pavia and Editore Ulrico Hoepli, Milano, 2003.

[56] K. H. Wiederkehr. Wilhelm Webers Stellung in der Entwicklung der Elektrizitätslehre. Dissertation, Hamburg, 1960.

[57] K. H. Wiederkehr. *Wilhelm Eduard Weber — Erforscher der Wellenbewegung und der Elektrizität (1804-1891)*, volume 32 of *Grosse Naturforscher*, H. Degen (ed.). Wissenschaftliche Verlagsgesellschaft, Stuttgart, 1967.

[58] K. H. Wiederkehr. Carl Friedrich Gauss (1777-1855) und Wilhelm Weber (1804-1891). In K. v. Meÿenn, editor, *Die Grossen Physiker*, Vol. 1, pages 357–370 and 522–524. Verlag C. H. Beck, München, 1997.

[59] F. Kirchner. Determination of the velocity of light from electromagnetic measurements according to W. Weber and R. Kohlrausch. *American Journal of Physics*, 25:623–629, 1957.

[60] L. Rosenfeld. The velocity of light and the evolution of electrodynamics. *Il Nuovo Cimento*, Supplement to vol. 4:1630–1669, 1957.

[61] A. E. Woodruff. The contributions of Hermann von Helmholtz to electrodynamics. *Isis*, 59:300–311, 1968.

[62] W. Weber and R. Kohlrausch. Über die Einführung absoluter elektrischer Maße. In S. Balke, H. Gericke, W. Hartner, G. Kerstein, F. Klemm, A. Portmann, H. Schimank, and K. Vogel, editors, *Ostwalds Klassiker der exakten Wissenschaften, new series, Vol. 5*. Friedrich-Vieweg & Sohn, Braunschweig, 1968. Commented by F. Kohlrausch and K. H. Wiederkehr.

[63] E. T. Whittaker. *A History of the Theories of Aether and Electricity*, volume 1: *The Classical Theories*. Humanities Press, New York, 1973.

[64] L. Rosenfeld. Kirchhoff, Gustav Robert. In C. C. Gillispie, editor, *Dictionary of Scientific Biography*, Vol. 7, pages 379–383, New York, 1981. Charles Scribner's Sons.

[65] A. E. Woodruff. Weber, Wilhelm Eduard. In C. C. Gillispie, editor, *Dictionary of Scientific Biography*, Vol. 14, pages 203–209, New York, 1981. Charles Scribner's Sons.

[66] K. Reich. *Carl Friedrich Gauß — 1777/1977*. Heinz Moos Verlag, München, 1977.

[67] M. N. Wise. German concepts of force, energy, and the electromagnetic ether: 1845–1880. In G. N. Cantor and M. J. S. Hodge, editors, *Conceptions of Ether — Studies in the History of Ether Theories 1740–1900*, pages 269–307, Cambridge, 1981. Cambridge University Press.

[68] P. M. Harman. *Energy, Force, and Matter – The Conceptual Development of Nineteenth-Century Physics*. Cambridge University Press, Cambridge, 1982.

[69] C. Jungnickel and R. McCormmach. *Intellectual Mastery of Nature — Theoretical Physics from Ohm to Einstein*, volume 1-2. University of Chicago Press, Chicago, 1986.

[70] W. A. Atherton. Gauss and Weber: An unlikely partnership. *Electronics and Wireless World*, 95:521–522, 1989.

[71] T. Archibald. Energy and the mathematization of electrodynamics in Germany, 1845-1875. *Archives Internationales d'Histoire des Sciences*, 39:276–308, 1989.

[72] K. H. Wiederkehr. Wilhelm Weber und die Entwicklung in der Geomagnetik und Elektrodynamik. In G. Berg, W. Eisenberg, W. Hergert, and H. Thieme, editors, *I. Weber - Symposium anläßlich des 100. Todestages W. Webers in Halle und Wittenberg am 20. und 21. Juni 1991*, pages 1–14. Fachbereich Physik der Martin-Luther-Universität Halle-Wittenberg und Wilhelm-Weber-Gesellschaft e. V., 1993.

[73] K. H. Wiederkehr. W. Weber und die Entwicklung der Elektrodynamik. In G. Berg, W. Eisenberg, W. Hergert, and W. Schreier, editors, *II. Weber - Symposium Die Gebrüder Weber - Wegbereiter interdisziplinärer Forshcung in Halle und Leipzig am 16. Oktober und 18. November 1993*, pages 39–54. Fachbereich Physik der Martin-Luther-Universität Halle-Wittenberg und Wilhelm-Weber-Gesellschaft e. V., 1993.

[74] K. H. Wiederkehr. Das Experiment von Wilhelm Weber und Rudolf Kohlrausch 1855 und Maxwells elektromagnetische Lichttheorie. In W. Schröder, editor, *The Earth and the Universe (A Festschrift in honour of Hans-Jürgen Treder)*, pages 452–463. Interdivisional Commission on History of the International Association of Geomagnetism and Aeronomy, Bremen-Rönnebeck, 1993.

[75] K. H. Wiederkehr. Wilhelm Weber und Maxwells elektromagnetische Lichttheorie. *Gesnerus, Part. 3/4*, 51:256–267, 1994.

[76] L. Hecht. Wilhelm Weber: giant of 19th century physics. *21st Century*, 10(2):53–54, 1997.

[77] O. Darrigol. *Electrodynamics from Ampère to Einstein*. Oxford University Press, Oxford, 2000.

[78] J. P. Wesley. Weber electrodynamics, Part I. General theory, steady current effects. *Foundations of Physics Letters*, 3:443–469, 1990.

[79] T. E. Phipps Jr. Toward modernization of Weber's force law. *Physics Essays*, 3:414–420, 1990.

[80] T. E. Phipps Jr. Weber-type laws of action-at-a-distance in modern physics. *Apeiron*, 8:8–14, 1990.

[81] J. P. Wesley. Weber electrodynamics, Part II. Unipolar induction, Z-antenna. *Foundations of Physics Letters*, 3:471–490, 1990.

[82] J. P. Wesley. Weber electrodynamics, Part III. Mechanics, gravitation. *Foundations of Physics Letters*, 3:581–605, 1990.

[83] J. P. Wesley. *Selected Topics in Advanced Fundamental Physics*. Benjamin Wesley Publisher, Blumberg, 1991.

[84] R. A. Clemente and A. K. T. Assis. Two-body problem for Weber-like interactions. *International Journal of Theoretical Physics*, 30:537–545, 1991.

[85] A. K. T. Assis. *Curso de Eletrodinâmica de Weber*. Setor de Publicações do Instituto de Física da Universidade Estadual de Campinas - Unicamp, Campinas, 1992. Notas de Física IFGW Número 5. Available online in PDF format at: http://www.ifi.unicamp.br/~assis/.

[86] T. E. Phipps Jr. Derivation of a modernized Weber force law. *Physics Essays*, 5:425–428, 1992.

[87] A. K. T. Assis and R. A. Clemente. The ultimate speed implied by theories of Weber's type. *International Journal of Theoretical Physics*, 31:1063–1073, 1992.

[88] S. Ragusa. Gravitation with a modified Weber force. *Foundations of Physics Letters*, 5:585–589, 1992.

[89] J. Guala-Valverde. Gravitational redshift revisited. *Galilean Electrodynamics*, 3:58, 1992.

[90] A. K. T. Assis and R. A. Clemente. The influence of temperature on gravitation. *Il Nuovo Cimento B*, 108:713–716, 1993.

[91] G. Galeczki. The ultimate speed and Weber's potential. *Physics Essays*, 6:448–450, 1993.

[92] A. K. T. Assis. *Eletrodinâmica de Weber - Teoria, Aplicações e Exercícios*. Editora da Universidade Estadual de Campinas - UNICAMP, Campinas, 1995. ISBN: 85-268-0358-1.

[93] G. Galeczki and P. Marquardt. Über die longitudinalen Ampèreschen Kräfte und Webers Elektrodynamik. *Fusion*, 16:15–17, 1995.

[94] J. Tennenbaum. An introduction to 'The significance of the 1845 Gauss-Weber correspondence'. *21st Century*, 9(3):2–5, 1996.

[95] L. Hecht. The significance of the 1845 Gauss-Weber correspondence. *21st Century*, 9(3):22–34, 1996.

[96] S. D'Agostino. Absolute systems of units and dimensions of physical quantities: a link between Weber's electrodynamics and Maxwell's electromagnetic theory of light. *Physis*, 33:5–51, 1996.

[97] L. Hecht. Experimental apparatus and instrumentation. *21st Century*, 9(3):35–40, 1996.

[98] T. E. Phipps, Jr. Clock rates in a machian universe. *Toth-Maatian Review*, 13:5910–5917, 1996.

[99] E. T. Kinzer and J. Fukai. Weber's force and Maxwell's equations. *Foundations of Physics Letters*, 9:457–461, 1996.

[100] F. Bunchaft and S. Carneiro. Weber-like interactions and energy conservation. *Foundations of Physics Letters*, 10:393–401, 1997.

[101] J. Fukai and E. T. Kinzer. Compatibility of Weber's force with Maxwell's equations. *Galilean Electrodynamics*, 8:53–55, 1997.

[102] A. K. T. Assis. *Mecânica Relacional*. Editora do Centro de Lógica, Epistemologia e História da Ciência da UNICAMP/FAPESP, Campinas, 1998. ISBN: 85-86497-01-0.

[103] Marcelo Bueno and A. K. T. Assis. *Cálculo de Indutância e de Força em Circuitos Elétricos*. Editora da UFSC/Editora da UEM, Florianópolis/Maringá, 1998. ISBN: 85-328-0119-6.

[104] J. Guala-Valverde. Gravitational mutual energy in Weber-type mechanics. *Physics Essays*, 11:164–165, 1998.

[105] A. K. T. Assis. *Uma Nova Física*. Editora Perspectiva, São Paulo, 1999. ISBN: 85-273-0199-7.

[106] V. F. Mikhailov. The action of an electrostatic potential on the electron mass. *Annales de la Fondation Louis de Broglie*, 24:161–169, 1999.

[107] A. K. T. Assis. *Relational Mechanics*. Apeiron, Montreal, 1999. ISBN: 0-9683689-2-1.

[108] A. G. Gluckman. On W. E. Weber's "Grundprincip der Elektrodynamik" and related earlier and contemporary studies. *Physics Essays*, 12:682–698, 1999.

[109] J. Guala-Valverde. *Inercia y Gravitacion.* Fundacion Julio Palacios, Neuquen, Argentina, 1999. In collaboration with J. Tramaglia and R. Rapacioli. Available at (accessed in 2007): http://www.educ.ar/educar/servlet/Downloads/-S_BD_LIBROSI/INERCIAYGRAVITACION.PDF.

[110] A. K. T. Assis and H. Torres Silva. Comparison between Weber's electrodynamics and classical electrodynamics. *Pramana Journal of Physics*, 55:393–404, 2000.

[111] A. K. T. Assis and J. Guala-Valverde. Mass in relational mechanics. *Apeiron*, 7:131–132, 2000.

[112] S. D'Agostino. *A History of the Ideas of Theoretical Physics.* Kluwer, Dordrecht, 2000.

[113] J. Guala-Valverde. A new theorem in relational mechanics. *Apeiron*, 8:132–138, 2001.

[114] M. d. A. Bueno and A. K. T. Assis. *Inductance and Force Calculations in Electrical Circuits.* Nova Science Publishers, Huntington, New York, 2001. ISBN: 1-56072-917-1.

[115] V. F. Mikhailov. Influence of an electrostatic potential on the inertial electron mass. *Annales de la Fondation Louis de Broglie*, 26:33–38, 2001.

[116] J. Guala-Valverde and P. Mazzoni. The unipolar dynamotor: a genuine relational engine. *Apeiron*, 8:41–52, 2001.

[117] R. A. Rapacioli. More about the claimed identity between inertial and gravitational mass. *Apeiron*, 8:1–9, 2001.

[118] J. P. Wesley. *Selected Topics in Scientific Physics.* Benjamin Wesley Publisher, Blumberg, 2002.

[119] A. K. T. Assis, K. Reich, and K. H. Wiederkehr. Gauss and Weber's creation of the absolute system of units in physics. *21st Century*, 15, No. 3:40–48, 2002.

[120] A. K. T. Assis and J. Guala-Valverde. On absolute and relative motions in physics. *Journal of New Energy*, 6:8–12, 2002.

[121] J. Guala-Valverde, P. Mazzoni, and R. Achilles. The homopolar motor: A true relativistic engine. *American Journal of Physics*, 70:1052–1055, 2002.

[122] A. K. T. Assis and J. Guala-Valverde. Frequency in relational mechanics. *Annales de la Fondation Louis de Broglie*, 28:83–97, 2003.

[123] V. F. Mikhailov. Influence of a field-less electrostatic potential on the inertial electron mass. *Annales de la Fondation Louis de Broglie*, 28:231–236, 2003.

[124] A. K. T. Assis and K. H. Wiederkehr. Weber quoting Maxwell. *Mitteilungen der Gauss-Gesellschaft*, 40:53–74, 2003.

[125] J. Fukai. *A Promenade Along Electrodynamics*. Vales Lake Publishing, Pueblo West, 2003.

[126] J. E. Junginger and Z. D. Popovic. An experimental investigation of the influence of an electrostatic potential on electron mass as predicted by Weber's force law. *Canadian Journal of Physics*, 82:731–735, 2004.

[127] K. H. Wiederkehr. Ein bisher unveröffentlichter Brief von Rudolf Kohlrausch an Andreas v. Ettingshausen von 1854, das Kohlrausch-Weber-Experiment von 1854/55 und die Lichtgeschwindigkeit in Wilhelm Webers Elektrodynamik. *NTM International Journal of History and Ethics of Natural Sciences, Technology and Medicine*, 12:129–145, 2004.

[128] D. F. Bartlett. Analogies between electricity and gravity. *Metrologia*, 41:S115–S124, 2004.

[129] A. K. T. Assis, K. Reich, and K. H. Wiederkehr. On the electromagnetic and electrostatic units of current and the meaning of the absolute system of units - For the 200th anniversary of Wilhelm Weber's birth. *Sudhoffs Archiv*, 88:10–31, 2004.

[130] J. Guala-Valverde. Distant matter in physics. *Annales de la Fondation Louis de Broglie*, 29:635–640, 2004.

[131] J. Guala-Valverde, R. Achilles, and R. Blas. Inertial mass: a changing entity? — Weber vs. Einstein, Weber plus Einstein or none? *Apeiron*, 12:351–352, 2005.

[132] J. P. Wesley and P. Marquardt. *Light a Photon Flux and Other Topics*. Benjamin Wesley Publisher, Blumberg, 2006.

[133] T. E. Phipps Jr. *Old Physics for New - a worldview alternative to Einstein's relativity theory*. Apeiron, Montreal, 2006.

[134] W. Weber. Vorwort bei der Übergabe der Abhandlung: *Elektrodynamische Maassbestimmungen, insbesondere Zurückführung der Stromintensitäts-Messungen auf mechanisches Maass*. Berichte über die Verhandlungen der Königl. Sächs. Gesellschaft der Wissenschaften zu Leipzig, mathematisch-physische Klasse, 17:55–61, 1855. Reprinted in Wilhelm Weber's *Werke*, Vol. 3, H. Weber (ed.), (Springer, Berlin, 1893), pp. 591-596.

[135] W. Weber and R. Kohlrausch. Über die Elektricitätsmenge, welche bei galvanischen Strömen durch den Querschnitt der Kette fliesst. *Annalen der Physik und Chemie*, J. C. Poggendoff (ed.), 99:10–25, 1856. Reprinted in Wilhelm Weber's *Werke*, Vol. 3, H. Weber (ed.), (Springer, Berlin, 1893), pp. 597-608.

[136] R. Kohlrausch and W. Weber. Elektrodynamische Maassbestimmungen insbesondere Zurückführung der Stromintensitäts-Messungen auf mechanisches Maass. *Abhandlungen der Königl. Sächs. Gesellschaft der Wissenschaften, mathematisch-physische Klasse*, 3:221–290, 1857. Reprinted in Wilhelm Weber's *Werke*, Vol. 3, H. Weber (ed.), (Springer, Berlin, 1893), pp. 609-676.

[137] W. Weber. Elektrodynamische Maassbestimmungen - über ein allgemeines Grundgesetz der elektrischen Wirkung. *Abhandlungen bei Begründung der Königl. Sächs. Gesellschaft der Wissenschaften am Tage der zweihundertjährigen Geburtstagfeier Leibnizen's herausgegeben von der Fürstl. Jablonowskischen Gesellschaft (Leipzig)*, pages 211–378, 1846. Reprinted in Wilhelm Weber's *Werke*, Vol. 3, H. Weber (ed.), (Springer, Berlin, 1893) pp. 25-214.

[138] G. T. Fechner. Ueber die verknüpfung der Faraday'schen inductionserscheinungen mit den Ampèreschen elektro-dynamischen erscheinungen. *Annalen der Physik*, 64:337–345, 1845.

[139] H. C. Oersted. Experiments on the effect of a current of electricity on the magnetic needle. *Annals of Philosophy*, 16:273–277, 1820.

[140] H. C. Oersted. Experiments on the effect of a current of electricity on the magnetic needle. In R. A. R. Tricker, *Early Electrodynamics — The First Law of Circulation*, pages 113–117, New York, 1965. Pergamon.

[141] A. M. Ampère. Sur les effets des courans électriques. *Annales de Chimie et de Physique*, 15:59–76, 1820. Available at: http://www.ampere.cnrs.fr/.

[142] A. M. Ampère. The effects of electric currents. In R. A. R. Tricker, *Early Electrodynamics - The First Law of Circulation*, pages 140–154, New York, 1965. Pergamon. Translated by O. M. Blunn.

[143] C. Blondel. *A.-M. Ampère el la Création de l'Électrodynamique (1820-1827)*. Bibliothèque Nationale, Paris, 1982.

[144] L. P. Williams. What were Ampère's earliest discoveries in electrodynamics? *Isis*, 74:492–508, 1983.

[145] P. Graneau. *Ampere-Neumann Electrodynamics of Metals*. Hadronic Press, Nonantum, 1985.

[146] J. R. Hofmann. Ampère, electrodynamics, and experimental evidence. *Osiris*, 3 (2nd Series):45–76, 1987.

[147] J. R. Hofmann. Ampère's invention of equilibrium apparatus: a response to experimental anomaly. *The British Journal for the History of Science*, 20:309–341, 1987.

[148] J. R. Hofmann. Ampère's electrodynamics and the acceptability of guiding assumptions. In A. Donovan et al., editor, *Scrutinizing Science*, pages 201–217. Kluwer Academic Publishers, Dordrecht, 1988.

[149] L. P. Williams. André-Marie Ampère. *Scientific American*, January:72–79, 1989.

[150] C. Blondel. Vision physique "éthérienne", mathématisation "laplacienne": l'électrodynamique d'Ampère. *Revue d'Histoire des Sciences*, 42:123–137, 1989.

[151] P. Graneau and N. Graneau. *Newton Versus Einstein – How Matter Interacts with Matter*. Carlton Press, New York, 1993.

[152] P. Graneau. *Ampere-Neumann Electrodynamics of Metals*. Hadronic Press, Palm Harbor, 2nd edition, 1994.

[153] J. R. Hofmann. *André-Marie Ampère, Enlightenment and Electrodynamics*. Cambridge University Press, Cambridge, 1996.

[154] P. Graneau and N. Graneau. *Newtonian Electrodynamics*. World Scientific, Singapore, 1996.

[155] T. E. Phipps, Jr. and J. Guala Valverde. Simulation of Ampèrian current elements by magnetic toroids. *21st Century*, 11:55–65, 1998.

[156] J. Lühr. Die Entstehung eines Demonstrationsexperiments: Zur Geschichte der Ampère'schen Stromwaage. In P. Heering, F. Rieß, and C. Sichau, editors, *Im Labor der Physikgeschichte — Zur Untersuchung historischer Experimentalpraxis*, pages 135–156. BIS der Carl von Ossietzky, Oldenburg, 2000.

[157] N. Graneau, T. Phipps Jr., and D. Roscoe. An experimental confirmation of longitudinal electrodynamic forces. *European Journal of Physics D*, 15:87–97, 2001.

[158] J. C. Maxwell. *A Treatise on Electricity and Magnetism*. Dover, New York, 1954.

[159] R. Clausius. Ueber die Vergleichung der elektrodynamische Grundgesetze mit der Erfahrung. *Annalen der Physik*, 10:608–618, 1880.

[160] F. Rohrlich. *Classical Charged Particles — Foundations of Their Theory*. Addison-Wesley, Reading, 1965.

[161] R. Skinner. *Mechanics*. Blaisdell, Waltham, 1969.

[162] J. M. Pearson and A. Kilambi. Velocity-dependent nuclear forces and Weber's electrodynamics. *American Journal of Physics*, 42:971–975, 1974.

[163] V. Bush. The force between moving charges. *Journal of Mathematical Physics*, 5:129–157, 1926.

[164] S. Parker. Electrostatics and current flow. *American Journal of Physics*, 38:720–723, 1970.

[165] R. W. Chabay and B. A. Sherwood. *Electric and Magnetic Interactions*. Wiley, New York, 1995.

[166] R. W. Chabay and B. A. Sherwood. *Matter & Interactions*, volume 2: Electric and Magnetic Interactions. Wiley, New York, 2002.

[167] W. G. V. Rosser. Magnitudes of surface charge distributions associated with electric current flow. *American Journal of Physics*, 38:265–266, 1970.

[168] W. K. H. Panofsky and M. Phillips. *Classical Electricity and Magnetism*. Addison-Wesley, Reading, 2nd edition, 1962.

[169] J. H. Poynting. On the connexion between electric current and the electric and magnetic inductions in the surrounding field. *Transactions of the Royal Society of London A*, 176:277–306, 1885.

[170] A. Marcus. The electric field associated with a steady current in long cylindrical conductor. *American Journal of Physics*, 9:225–226, 1941.

[171] R. W. Chabay and B. A. Sherwood. A unified treatment of electrostatics and circuits. Unpublished manuscript. Available at (accessed in 2007): http://www4.ncsu.edu/~rwchabay/mi/circuit.pdf.

[172] L. Bergmann and C. Schaefer. *Lehrbuch der Experimentalphysik*, volume 2: Elektrizitätslehre. Walter de Gruyter, Berlin, 1950.

[173] L. Bergmann and C. Schaefer. *Lehrbuch der Experimentalphysik*, volume 2: Elektrizität und Magnetismus. Walter de Gruyter, Berlin, seventh edition, 1987.

[174] O. Jefimenko. Demonstration of the electric fields of current-carrying conductors. *American Journal of Physics*, 30:19–21, 1962.

[175] O. D. Jefimenko. *Electricity and Magnetism*. Appleton-Century-Crofts, New York, 1966.

[176] O. D. Jefimenko. *Electricity and Magnetism*. Electret Scientific Company, Star City, 2nd edition, 1989.

[177] O. Jefimenko, T. L. Barnett, and W. H. Kelly. Confinement and shaping of electric fields by current-carrying conductors. *Proceedings of the West Virginia Academy of Science*, 34:163–167, 1962.

[178] R. Sansbury. Detection of a force between a charged metal foil and a current-carrying conductor. *Review of Scientific Instruments*, 56:415–417, 1985.

[179] D. F. Bartlett and S. Maglic. Test of an anomalous electromagnetic effect. *Review of Scientific Instruments*, 61:2637–2639, 1990.

[180] C. K. Whitney. Current elements in relativistic field theory. *Physics Letters A*, 128:232–234, 1988.

[181] H. Hayden. Possible explanation for the Edwards effect. *Galilean Electrodynamics*, 1:33–35, 1990.

[182] T. Ivezić. Electric fields from steady currents and unexplained electromagnetic experiments. *Physical Review A*, 44:2682–2685, 1991.

[183] R. C. Ritter and G. T. Gillies. Torsion balances, torsion pendulums, and related devices. *Review of Scientific Instruments*, 64:283–309, 1993.

[184] W. R. Moreau *et al.* Charge density in circuits. *American Journal of Physics*, 53:552–553, 1985.

[185] R. W. Chabay and B. A. Sherwood. Video demonstrating surface charge on a high-voltage circuit. Available at (accessed in 2007): http://www4.ncsu.edu/~rwchabay/mi/.

[186] D. F. Bartlett and B. F. L. Ward. Is an electron's charge independent of its velocity? *Physical Review D*, 16:3453–3458, 1977.

[187] C. S. Kenyon and W. F. Edwards. Test of current-dependent electric fields. *Physics Letters A*, 156:391–394, 1991.

[188] G. Bonnet. Electric field arising from a steady current passing through a superconductor. *Physics Letters A*, 82:465–467, 1981.

[189] J. C. Curé. A modified version of the Millikan oil drop experiment to test the probable existence of a new electrodynamic field. *Physics Letters B*, 116:158–160, 1982.

[190] J. P. Wesley. Weber electrodynamics extended to include radiation. *Speculations in Science and Technology*, 10:47–61, 1987.

[191] R. I. Gray. *Unified Physics*. Dahlgren, Virginia, 1988.

[192] D. F. Bartlett and W. F. Edwards. Invariance of charge to Lorentz transformation. *Physics Letters A*, 151:259–262, 1990.

[193] T. Ivesić. The relativistic electric fields arising from steady conduction currents. *Physics Letters A*, 144:427–431, 1990.

[194] N. Bilić. Comment on Lorentz invariance/noninvariance of the electric charge. *Physics Letters A*, 162:87–90, 1992.

[195] D. K. Lemon, W. F. Edwards, and C. S. Kenyon. Electric potentials associated with steady conduction currents in superconducting coils. *Physics Letters A*, 162:105–114, 1992.

[196] A. K. Singal. On the charge invariance and relativistic electric fields from a steady conduction current. *Physics Letters A*, 162:91–95, 1992.

[197] V. N. Strel'tsov. The electric field of a current-carrying conductor. *Hadronic Journal*, 15:457–461, 1992.

[198] G. G. Shishkin *et al.* Investigation of possible electric potential arising from a constant current through a superconductor coil. *Journal of Physics D*, 35:497–502, 2002.

[199] J. A. Hernandes and A. K. T. Assis. Electric potential due to an infinite conducting cylinder with internal or external point charge. *Journal of Electrostatics*, 63:1115–1131, 2005.

[200] J. C. Maxwell. *An Elementary Treatise on Electricity*. Clarendon Press, Oxford, 1881.

[201] M. A. Melehy. Thermal momentum in thermodynamics, part 2. interfacial electrification: a new consequence of the first and second laws. *Physics Essays*, 11:430–433, 1998.

[202] G. B. Arfken and H. J. Weber. *Mathematical Methods for Physicists*. Academic Press, San Diego, fourth edition, 1995.

[203] I. S. Gradshteyn and I. M. Ryzhik. *Table of Integrals, Series, and Products*. Academic Press, New York, 5th edition, 1994.

[204] R. N. Varney and L. H. Fisher. Electromotive force: Volta's forgotten concept. *American Journal of Physics*, 48:405–408, 1980.

[205] J. A. Hernandes, E. Capelas de Oliveira, and A. K. T. Assis. Potential, electric field and surface charges close to the battery for a resistive cylindrical shell carrying a steady longitudinal current. *Revista de la Facultad de Ingeniería (Chile)*, 12:13–20, 2004.

[206] M. A. Matzek and B. R. Russell. On the transverse electric field within a conductor carrying a steady current. *American Journal of Physics*, 36:905–907, 1968.

[207] W. R. Smythe. *Static and Dynamic Electricity*. McGraw-Hill, New York, 3rd edition, 1968.

[208] A. Sommerfeld. *Electrodynamics*. Academic Press, New York, 1964.

[209] W. G. V. Rosser. What makes an electric current "flow". *American Journal of Physics*, 31:884–885, 1963.

[210] W. R. Moreau. Charge distributions on DC circuits and Kirchhoff's laws. *European Journal of Physics*, 10:286–290, 1989.

[211] B.-S. Eylon and U. Ganiel. Macro-micro relationships: the missing link between electrostatics and electrodynamics in students' reasoning. *International Journal of Science Education*, 12:79–94, 1990.

[212] C. Schaefer. *Einführung in die theoretische Physik*, volume 3, part 1. Walter de Gruyter, Berlin, 1932.

[213] B. R. Russell. Surface charges and leaky shields. *American Journal of Physics*, 51:269–270, 1983.

[214] R. N. Varney and L. H. Fisher. Electric field associated with stationary currents. *American Journal of Physics*, 52:1097–1099, 1984.

[215] A. K. T. Assis and J. I. Cisneros. The problem of surface charges and fields in coaxial cables and its importance for relativistic physics. In F. Selleri, editor, *Open Questions in Relativistic Physics*, pages 177–185, Montreal, 1998. Apeiron.

[216] A. K. T. Assis and J. I. Cisneros. Surface charges and fields in a resistive coaxial cable carrying a constant current. *IEEE Transactions on Circuits and Systems I*, 47:63–66, 2000.

[217] G. F. Leal Ferreira and Mariângela T. de Figueiredo. A eletrostática na corrente elétrica em fios. *Revista Brasileira de Ensino de Física*, 25:374–377, 2003.

[218] J. A. Stratton. *Electromagnetic Theory*. McGraw-Hill, New York, 1941.

[219] A. K. T. Assis and A. J. Mania. Surface charges and electric field in a two-wire resistive transmission line. *Revista Brasileira de Ensino de Física*, 21:469–475, 1999.

[220] A. K. T. Assis, J. A. Hernandes, and J. E. Lamesa. Surface charges in conductor plates carrying constant currents. *Foundations of Physics*, 31:1501–1511, 2001.

[221] R. Welti. Las cargas superficiales y el flujo de energía en un circuito simple. *Revista Brasileira de Ensino de Física*, 27:577–582, 2005.

[222] J. A. Hernandes, E. Capelas de Oliveira, and A. K. T. Assis. Resistive plates carrying a steady current: electric potential and surface charges close to the battery. *Foundations of Physics Letters*, 18:275–289, 2005.

[223] J. A. Hernandes and A. K. T. Assis. The potential, electric field and surface charges for a resistive long straight strip carrying a steady current. *American Journal of Physics*, 71:938–942, 2003.

[224] P. Moon and D. E. Spencer. *Field Theory Handbook*. Springer, Berlin, 2nd edition, 1988.

[225] D. J. Griffiths and Y. Li. Charge density on a conducting needle. *American Journal of Physics*, 64:706–714, 1996.

[226] M. A. Heald. Electric fields and charges in elementary circuits. *American Journal of Physics*, 52:522–526, 1984.

[227] J. A. Hernandes and A. K. T. Assis. The electric field outside and inside a resistive spherical shell carrying a steady azimuthal current. *Physica Scripta*, 72:212–217, 2005.

[228] J. A. Hernandes and A. K. T. Assis. Electric potential for a resistive toroidal conductor carrying a steady azimuthal current. Physical Review E, Vol. 68, 046611, 2003.

[229] J. A. Hernandes and A. K. T. Assis. Surface charges and external electric field in a toroid carrying a steady current. *Brazilian Journal of Physics*, 34:1738–1744, 2004.

[230] C. Neumann. *Theorie der Elektricitäts- und Wärme-Vertheilung in einem Ringe*. Verlag der Buchhandlung des Waisenhauses, Halle, 1864.

[231] E. Weber. *Electromagnetic Theory – Static Fields and Their Mapping*. Dover, New York, 1965. Formerly titled: Electromagnetic Fields, Theory and Applications.

[232] H. Bateman. *Higher Transcendental Functions*, volume 1. McGraw-Hill, New York, 1953.

[233] E. Ley-Koo and A. Góngora T. Desarrollo armónico del potencial de Coulomb en coordenadas toroidales. *Revista Mexicana de Física*, 40:805–813, 1994.

[234] E. Weber. *Electromagnetic Fields: Theory and Applications*, volume 1. Wiley, New York, 1950.

[235] W. R. Smythe. *Static and Dynamic Electricity*. Hemisphere, New York, 3rd edition, 1989.

[236] P. M. Morse and H. Feshbach. *Methods of Theoretical Physics*, volume 2. McGraw-Hill, New York, 1953.

[237] D. E. Gray (editor). *American Institute of Physics Handbook*. McGraw-Hill, New York, third edition, 1972.

[238] C. Snow. Formulas for computing capacitance and inductance. *National Bureau of Standards Circular*, 544:1–69, 1954.

[239] P. Moon and D. E. Spencer. *Field Theory for Engineers.* Van Nostrand, Princeton, 1961.

[240] O. Jefimenko. Electric fields in conductors. *Physics Teacher*, 15:52–53, 1977.

[241] W. M. Saslow. Consider a spherical batery... *American Journal of Physics*, 62:495–501, 1994.

[242] J. M. Aguirregabiria, A. Hernandez, and M. Rivas. An example of surface-charge distribution on conductors carrying steady currents. *American Journal of Physics*, 60:138–141, 1992.

[243] J. M. Aguirregabiria, A. Hernandez, and M. Rivas. Surface charges and energy flow in a ring rotating in a magnetic field. *American Journal of Physics*, 64:892–895, 1996.

[244] H. Härtel. The electric voltage: What do students understand? What can be done to help for a better understanding? In W. Jung R. Duit and C. v. Rhöneck, editors, *Aspects of Understanding Electricity. Proceedings of an International Workshop*, pages 353–362, Kiel, 1985. IPN-Arbeitsberichte 59. Available at (accessed in 2007): http://www.astrophysik.uni-kiel.de/~hhaertel/PUB/voltage_IPN.pdf.

[245] H. Härtel. A qualitative approach to electricity. Institute for Research on Learning, Palo Alto, CA, Report No. IRL87-0001. A revised part of the full paper is available at (accessed in 2007): http://www.astrophysik.uni-kiel.de/~hhaertel/PUB/voltage_IRL.pdf, 1987.

[246] N. W. Preyer. Transient behavior of simple RC circuits. *American Journal of Physics*, 70:1187–1193, 2002.

[247] H. Torres S. and A. K. T. Assis. The influence of the electric field outside a resistive solenoid on the Aharonov-Bohm effect. *Revista de la Facultad de Ingenieria de la Universidad de Tarapaca (Chile)*, 9:29–34, 2001.

[248] D. Psillos, P. Koumaras, and A. Tiberghien. Voltage presented as a primary concept in an introductory teaching sequence on DC circuits. *International Journal of Science Education*, 10:29–43, 1988.

[249] A. Benseghir and J.-L. Closset. The electrostatics-electrokinetics transition: historical and educational difficulties. *International Journal of Science Education*, 18:179–191, 1996.

[250] B. A. Thacker, U. Ganiel, and D. Boys. Macroscopic phenomena and microscopic processes: Students understanding of transients in direct current electric circuits. *American Journal of Physics*, 67, Suppl. 1:S25–S31, 1999.

[251] A. P. French. *Newtonian Mechanics*. W. W. Norton & Company, New York, 1971.

[252] G. S. Ohm. The galvanic circuit investigated mathematically. In R. Taylor, editor, *Scientific Memoirs*, Vol. 2, pages 401–506, New York, 1966. Johnson Reprint Corporation. English translation by W. Francis.

[253] R. K. Pathria. *Statistical Mechanics*. Pergamon Press, Oxford, 1977.

[254] W. Weber. Elektrodynamische Maassbestimmungen insbesondere über elektrische Schwingungen. *Abh. sächs. Ges. Wiss.*, 6:105–241, 1864. Reprinted in Wilhelm Weber's *Werke*, Vol. 4, H. Weber (ed.), (Springer, Berlin, 1894), pp. 105-241.

[255] J. C. Poggendorff. Bermerkung zu dem Aufsatz des Herrn Prof. Kirchhoff. *Annalen der Physik*, 100:351–352, 1857. Reprinted in W. Weber's *Werke*, Vol. 4, p. 242, H. Weber (ed.), (Springer, Berlin, 1894).

[256] A. K. T. Assis. Arguments in favour of action at a distance. In A. E. Chubykalo, V. Pope, and R. Smirnov-Rueda, editors, *Instantaneous Action at a Distance in Modern Physics - "Pro" and "Contra"*, pages 45–56, Commack, 1999. Nova Science Publishers.

[257] J. A. Hernandes and A. K. T. Assis. Propagação de sinais em condutores com a eletrodinâmica de Weber e comparação com o eletromagnetismo clássico. In SBF, editor, *Anais do XXI Encontro Nacional de Física de Partículas e Campos, São Lourenço, 23 a 27/10/00*, page 1, São Paulo, 2000. Sociedade Brasileira de Física. Published at (accessed in 2007): http://www.sbf1.if.usp.br/eventos/enfpc/xxi/procs/res89/.

[258] J. A. Hernandes and A. K. T. Assis. Propagação de sinais em condutores segundo a eletrodinâmica de Weber. *Ciência e Natura*, 23:7–26, 2001.

[259] S. Schaffer. Accurate measurement is an english science. In M. Norton Wise, editor, *The Values of Precision*, chapter 6, pages 137–172. Princeton University Press, Princeton, 1995.

Index

Aharonov-Bohm effect, 192, 193
Ampère's
  force between current elements, 15, 16
  law, 16, 19
Ampère, A. M., 12, 16
Anomalous electromagnetic effect, 36, 37
Arfken, G. B., 158
Assis, A. K. T., 8, 12, 34, 62

Barnett, T. L., 34, 42, 88, 90, 91, 129–131, 185, 193
Bartlett, D. F., 36–40, 43, 87
Becquerel, 66
Bergmann, L., 30, 31, 34, 36, 42, 88–90, 110, 111, 116, 193
Bessel
  equation, 53
  function, 48, 56
Beuzenberg, 40, 42
Biot-Savart's law, 19

Capacitance, 81, 179, 182
Cavendish, H., 66
Chabay, R. W., 25, 31, 41, 42, 66, 81, 149, 191, 192
Charge density
  angular, 203
  linear, 56, 57, 59, 79, 126, 178, 182, 200, 203
  surface, 8, 11, 21, 24, 26, 27, 40–42, 59, 68, 77, 79–82, 85, 86, 89, 94, 107, 108, 113, 115, 126–128, 141, 142, 145, 146, 153, 158–160, 176–179, 199, 200, 203, 208, 209, 211, 214, 215
  volume, 8–13, 30, 85, 86, 104, 196, 203, 213
Christy, R. W., 9
Classical electrodynamics, 3, 20, 22, 26, 84
Clausius's
  force, 22, 82
  postulate, 9
Clausius, R., 9, 10, 17–19
Closed circuit, 4, 9, 10, 17–20, 31, 41, 68, 76, 82–84, 157, 191, 196, 199, 208, 214
Conservation of charge, 215
Coombes, C. A., 11, 81
Corson, D. R., 11
Coulomb's
  force, 15, 30, 66–68, 195, 197, 209, 210
  law, 14, 198, 199
Current element, 9, 12, 15, 16

Davy, H., 66
Dirac delta function, 48
Dirichlet boundary condition, 47
Displacement current, 216
Du Mont power supply, 32

Edwards, W. F., 10, 43
Electromagnetic signal, 211, 216
Electromotive force, 3, 10, 18, 23, 42, 43, 67, 68, 157, 192, 193, 196–200, 202, 205–211, 213
Elliptic-cylindrical coordinates, 124
Emf, 23, 29, 30, 36, 42, 45, 67, 68, 76, 82, 83, 87, 142, 143,

236

193, 196
English translation, 12, 14, 15, 17, 24, 195, 209, 211, 213
Equipotential lines, 30, 34, 35, 88–91, 97, 100, 105, 110, 111, 115–117, 119, 121, 128–131, 139, 140, 147, 158, 159, 170, 171, 183–185, 187, 193
Euler gamma, 175
Euler-Mascheroni constant, 57

Faraday cage, 38, 39, 43
Faraday's law of induction, 15
Fechner's hypothesis, 15, 16, 210
Fechner, G. T., 15, 19
Feynman, R. P., 13, 14, 27
Field
    electric, 3–5, 8–14, 17, 18, 20–32, 34–36, 40–43, 46, 47, 51, 54, 55, 59, 65–67, 69, 75–77, 79–91, 93–97, 99–101, 103, 104, 106, 107, 110–113, 115, 117–123, 125–130, 139–143, 148, 151, 153, 156–160, 163, 168, 173–177, 181, 184–188, 192, 193, 196–203, 205, 206, 208, 210, 211
    electromagnetic, 26
    electrostatic, 21, 67, 199
    Liénard-Wiechert, 10
    magnetic, 4, 8–10, 12–14, 26, 27, 31, 77, 82, 84, 85, 93, 100, 104, 126, 191, 192
Flux of energy, 26
Force
    electromagnetic, 210, 211
    Liénard-Schwarzschild, 82
    magnetic, 12, 84, 85, 109, 110, 126, 127, 186
Fourier
    series, 50, 139, 143–146, 154, 159, 167, 169, 176, 186
    transform, 50
Fourier, J. B. J., 196
French, A. P., 194

Galvanic
    circuit, 195, 196, 199, 209
    current, 17, 18, 195
Ganiel, U., 41, 42
Gauss's
    force, 22
    law, 15, 51, 55, 69, 85, 89, 95, 96, 115, 126, 141, 144, 157, 174, 203
Gibbs phenomenon, 167
Green function, 47–50, 52, 53
Greenwich Meridian, 151
Griffiths, D. J., 11, 137

Härtel, H., vi, 192
Hall
    effect, 16, 67, 77, 82, 84, 85, 104, 186
    electric field, 126
    voltage, 84
Hamburg University, iii
Heald, M. A., 137, 140, 146, 192
Hernandes, J. A., 37
Humboldt Foundation, iii

Inertial frame of reference, 4, 20, 45, 65
International System of Units SI, 4, 203, 213

Jackson, J. D., 9, 10, 14, 46, 50, 56, 155, 163, 183, 191, 192
Jefimenko, O., 13, 14, 31, 32, 34, 36, 42, 67, 88–91, 110–113, 116, 117, 122, 128–131, 137, 184, 185, 187, 191–193

Kelly, W. H., 34, 42, 88, 90, 91, 129–131, 185, 193
Kenyon, C. S., 10, 43
Keplerian elliptical orbit, 210
Kilambi, A., 19
Kirchhoff, G., 4, 12, 24, 77, 90, 192, 209, 211, 213–216
Kohlrausch, R., 15, 209, 211, 216
Kronecker's delta funcion, 154, 169

Laplace's equation, 39, 69, 80, 94, 95, 123, 125, 138, 153, 166, 168, 203, 213
Laue, H., 11, 81
Legendre
   equation, 153, 166
   function, 153, 155, 156, 166, 167, 175, 182
   polynomial, 153, 154, 166
Leighton, R. B., 14, 27
Lemon, D. K., 10, 43
Liénard-Schwarzschild's force, 82
Liénard-Wiechert
   field, 10
Light velocity, 10, 29, 66, 213, 215, 216
Lines of electric field, 30, 31, 34, 35, 87–91, 95, 97, 110–112, 115, 117, 120, 122, 128–130, 139–141, 148, 158, 159, 184, 185, 187, 193
Lorentz's
   force, 22, 82–84
   transformation, 14
Lorrain, F., 11
Lorrain, P., 11

Maglic, S., 36–40, 87
Magnetic
   flux, 191
   vector potential, 193, 214
Magnetic circuital law, 84
Matzek, M. A., 84
Maxwell, J. C., 17–19, 46, 216
Melehy, M. A., 46
Method of images, 3, 36, 38, 39, 43, 45, 46, 104, 107, 192
Milford, F. J., 9
Moreau, W. R., 40, 42, 196
Motional electric field, 22, 83, 85, 96

Neumann, C., 164
Newton's second law of motion, 186
Newtonian
   force, 68, 210
   Mechanics, 194

Non-Coulomb force, 67, 68, 196

Ohm's law, 8, 9, 24, 26, 66, 79, 89, 94, 98, 103, 109, 123, 137, 152, 157, 168, 195, 196, 209, 210, 213–215
Ohm, G. S., 196, 203, 206–209, 213
Ohmic
   conductor, 163, 164, 168
   resistance, 81, 113
   toroid, 167

Parker, S., 25, 35, 36, 40, 42, 67
Pearson, J. M., 19
Poggendorff, J. C., 211, 216
Poisson's equation, 15, 47
Poisson, S. D., 197
Popovic, B. D., 11, 12
Potential difference, 3, 9, 10, 23, 29, 30, 43, 45, 65, 67, 68, 76, 81–83, 86, 90, 96, 141, 151, 186, 196, 210
Poynting's vector, 26, 27
Poynting, J. H., 26
Preyer, N. W., 192
Principle of superposition, 65
Purcell, E. M., 13, 14

Ratio of electromagnetic and electrostatic units of charge, 15
Reitz, J. R., 9
Relativistic effect, 13, 14
Riecke, 18
Riemann, B., 10, 22
Ritz, W., 10, 16, 22
Rohrlich, F., 19
Rosser, W. G. V., 26, 84, 191
Russell's theorem, 68, 135
Russell, B. R., 9, 68, 84, 94, 101
Ryan, 40, 42

Sands, M., 14, 27
Sansbury, R., 30, 36, 37, 39, 40, 42, 86, 88, 89
Saslow, W. M., 191
Schaefer, C., 30, 31, 34, 36, 42, 67, 88–90, 110, 111, 116, 193

Second order
    components of Weber's force, 15
    correction, 66
    effect, 10, 21, 87
    electric field, 10, 12, 20–22, 29, 42, 43, 83, 96, 193
    force, 20–22, 83, 84, 86, 193
Seely, S., 11
Sherwood, B. A., 25, 31, 41, 42, 66, 81, 149, 191, 192
Skinner, R., 19
Smythe, W. R., 84
Sommerfeld, A., 87, 110
State University of Campinas - UNICAMP, iii
Stratton, J. A., 103
Superconductor, 43, 193
Syme, 40, 42

Telegraphy equation, 4, 12, 192, 211, 213, 215, 216
Toroidal coordinates, 164–166, 173, 179, 183

Van de Graaff generator, 32, 67
Volta, A. G. A. A., 67
Voltage, 3, 7, 8, 10, 12, 17, 18, 20–23, 25, 30, 32, 36, 40–43, 45, 63, 68, 76, 86, 87, 89, 98, 142, 143, 193

Ward, B. F. L., 43
Weber's
    constant, 216
    electrodynamics, 10, 12, 14, 16–20, 22, 82, 83
    force, 14–17, 19, 82, 83, 209, 213, 214
    law, iii, 14–17, 19, 20, 43
Weber, H. J., 158
Weber, W. E., i, 4, 13–17, 19, 24, 68, 75, 77, 79, 90, 163, 176–178, 188, 191, 192, 194–211, 213–216
Weizmann Institute, 41
Whittaker, E. T., 18, 19

Wiederkehr, K. H., 14
Wronskian, 49, 58

Zeroth order
    effect, 21
    electric field, 20–22, 29, 43, 46, 51, 59, 87, 192, 193
    force, 10, 21, 22, 29, 36, 42, 51, 52, 55, 57, 59, 61, 63, 86